Life Support

Life Support

The Environment and Human Health

edited by Michael McCally

The MIT Press
Cambridge, Massachusetts
London, England

This book was set in Bembo by Achorn Graphic Services Inc.

Printed and bound in the United States of America.

Library of Congress Cataloging-in-Publication Data

Life support : the environment and human health / edited by Michael McCally.
p. cm.
Includes bibliographical references and index.
ISBN 13: 978-0-262-13414-9 (alk. paper) 978-0-262-63257-7 (pbk: alk. paper);
ISBN 10: 0-262-13414-4 (alk. paper) 0-262-63257-8 (pbk: alk. paper)
1. Environmental health. 2. Public health—Environmental aspects. I. McCally, Michael.

RA565 .L54 2002
615.9′02—dc21

2002020191

10 9 8 7 6 5 4 3

Contents

Preface vii
Contributors xi

1
Environment, Health, and Risk 1
Michael McCally

2
Urban and Transboundary Air Pollution 15
David C. Christiani and Mark A. Woodin

3
Water Quality and Water Resources 39
John M. Balbus

4
Human Health and Heavy Metals Exposure 65
Howard Hu

5
Population, Consumption, and Human Health 83
J. Joseph Speidel

6
Global Climate Change and Health 99
Andrew Haines, Anthony J. McMichael, and Paul R. Epstein

7
Species Loss and Ecosystem Disruption 119
Eric Chivian

8
Ozone Depletion and Ultraviolet Radiation 135
Frank R. de Gruijl and Jan C. van der Leun

9
Environmental Endocrine Disruption 147
Gina M. Solomon and Ted Schettler

10
Body Burdens of Industrial Chemicals in the General Population 163
Joe Thornton, Michael McCally, and Jeff Howard

11
Cancer and the Environment 201
Richard W. Clapp

12
Radiation and Health 211
Maureen Hatch and Michael McCally

13
The Science of Risk Assessment 231
John C. Bailar III and A. John Bailer

14
The Precautionary Principle 239
Ted Schettler, Katherine Barrett, and Carolyn Raffensperger

15
Vulnerable Populations 257
Philip Landrigan and Anjali Garg

16
War and the Environment 273
Jennifer Leaning

17
Sustainable Health Care and Emerging Ethical Responsibilities 285
Andrew Jameton and Jessica Pierce

Index 297

Preface

Life Support: The Environment and Human Health builds on the book *Critical Condition: Human Health and the Environment,* with new material and a focus on solutions. Ten new chapters have been added, and the others have been substantially revised. I have asked contributors to conclude their essay with specific recommendations for action—prescriptions that, in their opinion, are necessary for solution of the environmental health problems discussed in their chapter.

Edited books may represent diverse points of view, and the contributors' essays speak for themselves. But the authors share several beliefs that are the theme of this book. Humankind's impact on the earth has reached such magnitude and environmental degradation has proceeded to such a point that we are well into a global environmental crisis. *Crisis* is an appropriate and useful word in this instance. The specifications of the crisis are spelled out in each of this book's seventeen chapters. It is a frightening story, no doubt about it. At the same time, we all share an optimism that good ideas—some suggested here—energy, and luck can help our grandchildren achieve effective environmental protection and sustainability.

In addition, the authors of this book are all health professionals and share the belief that we cannot continue working to improve human health and health care while ignoring the collapsing ecosystems that sustain us. The concepts of health and the environment are deeply embedded, one in the other.

Life Support has a particular history. In 1991, as preparations were being concluded for the United Nations Conference on Environment and Development (UNCED) in Rio de Janeiro, it was clear to four colleagues—Eric Chivian, Andrew Haines, Howard Hu, and myself—that the health dimensions of environmental degradation had been neglected. Human health was not on the Rio agenda, even though health is certainly determined by and reflects the state of the environment. A message about the connections of health and the environment needed to be taken to Rio.

In fact, it seemed then that little attention was given to health in any environmental discussion. There appeared to be little connection between the public health community and the environmental movement. No medical or public health organization worked on environmental issues. The staffs of mainstream environmental organizations had no medical or health expertise. Community groups complained that they could not find physicians or health experts willing to work with them on environmental health issues. The Environmental Protection Agency, with over 20,000 employees, had fewer than a half dozen physician employees.

In late 1991, Eric Chivian and I met with Noel Brown, then Director for North America of the United Nations Environmental Programme. Out of these meetings grew the notion for a book, to be called *Critical Condition*. With our two colleagues, we rushed to assemble a text, and with Chivian's leadership had it done in time for the Rio meeting.

Now it is ten years later and some things are different. While the EPA still has fewer than six physicians on staff, the connections between human health and the environment are now more widely appreciated. Environmental issues like global warming are presented in terms of their consequences for human health. Health professionals serve on the staff of many environmental organizations. Active national and international coalitions do health-oriented work on toxics, climate change, biodiversity, and habitat destruction. New organizations, like Second Nature and the Center for Health and the Global Environment at the Harvard Medical School, do research, advocacy, and training. Some observers believe that we can see the outlines of an international environmental health movement.

I took on the responsibility to edit this new volume with three objectives: to update the original work, to expand the coverage, and to focus on solutions or prescriptions. I turned to colleagues who I felt were most expert in the major topics of environmental health and who could write chapters accessible to informed lay readers, as well as trainees and professionals. In addition to the topics traditionally covered in standard texts—air, food, and water quality, together with occupational health—*Life Support* deals with the new public threats: climate change, stratospheric ozone depletion, habitat loss, war and other military activity, and resource scarcity. Environmental health science is developing rapidly. The endocrine-disrupter hypothesis has given a new vitality and complexity to toxicology. In policy science, the multidisciplinary craft of risk assessment has been augmented by new notions of the precautionary principle.

As in the earlier book, the important roles of physicians and health professionals are stressed. Health-trained professionals ought to become central figures in environmental policy discussions. I do not believe that this is a matter of professional chauvinism. In periods of crisis, humans naturally think of their own health and safety. I believe that citizens' environmental awareness and, more important, willingness to act in sustainable and environmentally protective ways will come through the perception of threats to human health, damage to the health of children in particular. It is my experience from public speaking and legislative testimony that arguments for the environment made from the perspective of human health are persuasive.

An edited volume involves collaboration. I want to thank the authors for their expertise, writing skills, good humor, and, perhaps most important, for finishing their work on schedule. Nine of the essays in this volume were written for the *Canadian Medical Association Journal,* and I want to thank *CMAJ* Editor Dr. John Hoey both for conceiving the Human Health and the Environment Series in his journal and for permission to reprint revisions of these essays here. Clay Morgan of The MIT Press was enthusiastic about this project from the beginning and saw it skillfully to conclusion. Carol Capello managed the preparation of the final manuscripts and helped significantly in the editing.

I am indebted to and thank many colleagues for my continuing education in the science and policy of environmental health: Philip Landrigan, Eric Chivian, Andrew Haines, Andrew Jameton, Paul Epstein, Howard Frumkin, Ted Schettler, Joe Thornton, Ross Vincent, Paul Ehrlich, Anne Ehrlich, Peter Orris, Robert Musil, Tony McMichael, Sandra

Steingraber, Michael Lerner, Gary Cohen, Doris Cellarius, Charlotte Brody, John Last, Jill Stein, Barbara Sattler, Jack Weinberg, Christine Cassel, and Robert Lawrence.

Life Support is dedicated to my granddaughter, Micah McCally, and her contemporaries, with the hope that current efforts to pass along a healthy planet may succeed.

Physicians for Social Responsibility (PSR) is the voice of the medical profession in the growing United States and world environmental health movement. PSR speaks with both moral authority and scientific expertise, and I am pleased to acknowledge PSR's leadership. Its programs on toxics, air and water quality, and global warming are scientifically sound and effective. This volume continues the tradition of an earlier PSR text from MIT Press: *Critical Condition: Human Health and the Environment.* The royalties from *Life Support* will go to Physicians for Social Responsibility.

Contributors

John C. Bailar III, M.D., Ph.D.
Professor Emeritus
Department of Health Studies
University of Chicago
Chicago, Ill.
jcbailar@midway.uchicago.edu

A. John Bailer, Ph.D.
Professor of Statistics
Department of Mathematics and Statistics
Miami University
Oxford, Ohio
and the Education and Information Division
National Institute for Occupational Safety and Health
Cincinnati, Ohio
ajbailer@muohio.edu

John M. Balbus, M.D., M.P.H.
Director, Center for Risk Science and Public Health
Associate Professor of Environmental and Occupational Health and of Medicine
The George Washington University
Washington, D.C.
eohjmb@gwumc.edu

Katherine Barrett, Ph.D.
Research Associate
Polis Project on Ecological Governance
Project Director
Science and Environmental Health Network
University of Victoria
Victoria, Canada

Eric Chivian, M.D.
Director
Center for Health and the Global Environment
Harvard Medical School
Boston, Mass.
eric_chivian@hms.harvard.edu

David C. Christiani, M.D., M.P.H., M.S.
Professor of Epidemiology and Occupational Medicine
Department of Environmental Health
Harvard School of Public Health
Boston, Mass.
dchris@hohp.harvard.edu

Richard W. Clapp, M.P.H., D.Sc.

Associate Professor
Department of Environmental Health
Boston University School of Public Health
Boston, Mass.
Rclapp@bu.edu

Paul R. Epstein

Associate Director
Center for Health and the Global Environment
Harvard Medical School
Boston, Mass.
paul_epstein@hms.harvard.edu

Anjali Garg, M.S.

Health Policy Analyst
Center for Children's Health and the Environment
Department of Community and Preventive Medicine
Mount Sinai School of Medicine
New York, NY
anjali.garg@mssm.edu

Frank R. de Gruijl, Ph.D.

Associate Research Professor
Department of Dermatology
Leiden University Medical Center
Leiden, The Netherlands

Maureen Hatch, Ph.D.

Associate Professor
Department of Community and Preventive Medicine
Mount Sinai School of Medicine
New York, N.Y.
maureen.hatch@mssm.edu

Andrew Haines, M.B., B.S., M.D.

Dean and Professor of Public Health and Primary Care
London School of Hygiene and Tropical Medicine
London, England
andy.haines@lshtm.ac.uk

Jeff Howard, M.S.

Department of Science and Technology Studies
Rensselaer Polytechnic Institute
Troy, N.Y.
howarjrpi.edu

Howard Hu, M.D., M.P.H., Sc.D.

Associate Professor of Occupational and Environmental Medicine
Department of Environmental Health
Harvard School of Public Health
Associate Professor of Medicine
Channing Laboratory
Department of Medicine
Harvard Medical School
Boston, Mass.
hhu@hsph.harvard.edu

Andrew Jameton, Ph.D.

Professor
Department of Preventive and Societal Medicine
University of Nebraska Medical Center
Omaha, Neb.
ajameton@UNMC.edu

Philip Landrigan, M.D., MSc.

Professor of Pediatrics
Director, Center for Children's Health and the Environment
Chairman
Department of Community and Preventive Medicine
Mount Sinai School of Medicine
New York, N.Y.
phil.landrigan@mssm.edu

Jennifer Leaning, M.D., S.M.H.

Professor of International Health
Director, Program on Humanitarian Crises
FXB Center for Health and Human Rights
Harvard School of Public Health
Boston, Mass.
jleaning@hsph.harvard.edu

Jan C. van der Leun, Ph.D.
Professor Emeritus
Department of Dermatology
University of Utrecht
Guest Senior Scientist/Ecofys, bv.
Utrecht, The Netherlands

Michael McCally, M.D., Ph.D.
Professor
Department of Public Health and Preventive Medicine
Oregon Health Sciences University
Portland, Oregon
mccallym@ohsu.edu

Anthony J. McMichael, MBBS, PhD
Director
National Centre for Epidemiology and Population Health
Australian National University
Canberra, Australia
tony.mcmichael@anu.edu.au

Jessica Pierce, Ph.D.
Visiting Fellow, Center for Values and Social Policy
Department of Philosophy
University of Colorado at Boulder
Boulder, Colo.
piercejessica@usa.net

Carolyn Raffensperger, M.A., J.D.
Executive Director
Science and Environmental Health Network
Ames, Iowa
raffensperger@cs.com

Ted Schettler, M.D., M.P.H.
Science Director
Science and Environmental Health Network and
Department of Medicine
Boston Medical Center
Boston, Mass.
tschettler@igc.org

Gina M. Solomon, M.D., M.P.H.
Senior Scientist
Natural Resources Defense Council
and
Assistant Professor of Medicine
Department of Medicine
University of California at San Francisco
San Francisco, Calif.
gsolomon@nrdc.org

J. Joseph Speidel, M.D., M.P.H.
Program Director for Population
William and Flora Hewlett Foundation
Menlo Park, Calif.
jspeidel@hewlett.org

Joe Thornton, Ph.D.
Research Scientist
The Earth Institute and
Department of Biological Sciences
Columbia University
New York, N.Y.
jt121@columbia.edu

Mark A. Woodin, Sc.D., M.S.
Lecturer
Department of Civil and Environmental Engineering and Adjunct Assistant Professor
Department of Family Medicine and Community Health
School of Medicine
Tufts University
Medford, Mass.
mark.woodin@tufts.edu

Environment, Health, and Risk

Michael McCally

1

At least since the time of Hippocrates's essay "Air, Water, Places," humans have been aware of the many connections between health and the environment.[1] Improved water, food, and milk sanitation, reduced physical crowding, improved nutrition, and central heating with cleaner fuels were the developments most responsible for the great improvements in public health achieved during the twentieth century. But prior environmental health problems, local in their effects and short in duration, have changed dramatically within the last 25 years. Today's problems are persistent and global. Together, global warming, population growth, habitat destruction, and resource depletion have produced a widely acknowledged environmental crisis.[2–5] These long-term environmental problems are not amenable to quick technical fixes and will require profound social changes for their solution.

The Environmental Crisis

For the first time, human beings are altering the basic operations of the earth's atmosphere, geosphere, and biosphere. Four prominent biologists have noted with concern that "human alteration of Earth is substantial and growing. Between one-third and one-half of the land surface has been transformed by human action; the carbon dioxide concentration in the atmosphere has increased nearly 30 percent since the beginning of the Industrial Revolution; more atmospheric nitrogen is fixed by humanity than by all natural terrestrial sources combined; more than half of all accessible surface fresh water is put to use by humanity, and about one-quarter of bird species on Earth have been driven to extinction."[6] In 1992, the World Scientists' Warning to Humanity was endorsed by more than 1,600 scientists from 70 countries, among them 104 Nobel laureates, including most of the science-prize recipients. The warning cited clear evidence of a growing environmental crisis.[7]

Biologists have observed startling declines in frog populations around the world, even in isolated and relatively pristine environments. It appears that no single factor is responsible. Rather, the health and reproductive success of amphibians is being damaged by interactions between an increase in the intensity of ultraviolet (UV) light (because of thinning of stratospheric ozone), traces of globally distributed toxic chemicals, competition from introduced predator species, and infections caused by virulent fungi and bacteria.[8] The declining health of frogs, birds, and thousands of other organisms may be the clearest indication of environmental threats to human health. Although local environmental tragedies of climate change, species extinction, and deforestation have marked every period of human history, today's environmental degradation is rapidly creating an unprecedented global crisis. The driving forces are population growth and industrialization.[9–11]

The rate of industrialization, which has far outpaced the growth in population, is a powerful determinant of environmental transformation. In the past 100 years, the world's industrial production increased 100-fold.[12] From 1950 to 1999, manufacturing increased by factor of 7, the use of fossil fuels by a factor of 4, the number of automobiles by a factor of 10 (from 53 million to 520 million), and the production of synthetic chemicals by several orders of magnitude.[13] The impact on the global environment has been dramatic. In 150 years, human activity has increased the atmospheric concentration of carbon dioxide by nearly 30 percent, doubled the concentration of methane (both potent greenhouse gases), and introduced long-lived ozone-destroying chlorofluorocarbons into the stratosphere. Human-made emissions of sulfur and nitrogen, which led to acid rain, now equal or exceed the natural flux of these elements. And human-made emissions of lead, cadmium, and zinc exceed the natural flux by a factor 18.[12]

Human and Ecosystem Health

From a medical standpoint, health is viewed as an attribute of the individual. The fields of medicine and public health have traditionally acknowledged environmental causes of illness and assigned risk to specific exposures. In the past decade, biologists, ecologists, and physicians have also developed a concept of ecosystem health.[2,14,15] This idea recognizes that humans are participants in complex ecosystems and that their potential for health is proportional to the health function of those ecosystems. An ecosystem-based health perspective takes into account the health-related services that the natural environment provides (e.g., soil production, pollination, and water cleansing) and acknowledges the fundamental connection between an intact environment and human health.[14] An ecosystem health stance is a nonanthropocentric, holistic worldview increasingly shared by biological scientists.[15]

Fundamental human beliefs have produced this environmental transformation. In the anthropocentric view of the world, humans are the most important of all the species and should have dominion over nature. Many of us believe that any harm we do we can undo with ingenuity and technology and that our individual and collective impact on the environment will result in only imperceptible changes.

Global environmental issues at this "special moment in history," as Bill McKibben has termed the turn of the millenium, are unique in their scope and consequences, and discussion of them may be emotionally and politically charged.[16] Global change may seem so remote from our daily lives that we become indifferent to the environmental apocalypse we face. We may not perceive the actual degradation of the earth. Some of us distance ourselves from the discussion because we find it frightening or overwhelming. As a result, policymakers and politicians are not pressed to confront the consequences of the continuing expansion of human enterprise. For example, climate change produced by the accumulation of greenhouse gases, primarily carbon dioxide, is in process. The 1995 second report of the Intergovernmental Panel on Climate Change (IPCC) concluded that "man's impact on climate is discernible," and many climate scientists believe that we are already experiencing global warming effects.[17] More recently, the IPCC and the U.S. National Academy of

Sciences have both concluded that the earth is warming and that greenhouse gases are part of the cause. But it is difficult for us to acknowledge that industrial carbon dioxide—invisible, odorless, and nontoxic—is a pollutant and a waste that we release into the atmosphere. International political leadership has only begun to seek solutions to the global issues of climate change, toxic pollution, species loss, and deforestation.

Health and the Environment

The problems resulting from environmental change pose new challenges for traditional public health science.[18] The health effects of global change are often indirect and difficult to assess, and the quality of evidence for the health-related outcomes of global environmental change varies widely.[3] For example, the prevalence of malaria has increased worldwide, but no clear relation to climate change has been established. Similarly, exposure to UV light (especially UVB) increases skin cancer and cataract formation, but large studies across geographic areas with different levels of UV exposure have not been performed. Furthermore, the health science necessary to understand global environmental change is increasingly interdisciplinary and requires collaboration over long periods among meteorologists, chemists, biologists, agronomists, and health scientists. Organizing and funding such science is difficult. Finally, the science of global change frequently relies on computer models to suggest the direction and magnitude of change, but politicians and policymakers are loath to commit resources to predicted but unproven future outcomes.

Environmental degradation exaggerates the imbalance between population and resources, increases the costs of development, and worsens the extent and severity of poverty. Population growth and the corporatization of agriculture and forestry have forced poor people onto land that is the least productive and ecologically the most fragile. In crowded or poor countrysides, people often abandon traditional and sustainable land-use practices in favor of short-term survival strategies, such as farming on steep slopes and living in areas threatened by flood or drought. The need for farmland, fuelwood, and timber for export results in deforestation, which, in turn, increases soil erosion, flooding, and mud slides and reduces agricultural productivity. In short, interactions between poverty, population growth, and environmental degradation impede sustainable economic development and worsen population health.[19]

It is important for scientists to anticipate the potential consequences of environmental change.[20] Serious environmental problems are often unknown or unrecognized. The stratospheric ozone hole produced by chlorofluorocarbons, although anticipated, was discovered by accident.[21] At the time of the first major international conference on the environment, held in Stockholm in 1972, global warming, acid rain, and tropical deforestation were not recognized as major problems. Explanations of the decline in amphibian populations, cancer outbreaks in fish, and the bleaching of coral reefs are still inadequate today. Furthermore, change in natural systems may be sudden and nonlinear. For example, fish populations that have remained stable during long periods of intense harvesting may suddenly collapse.

Two recent developments have drawn renewed attention to the health risks of industrial chemicals known as persistent organic pollutants (POPs): the identification of medical waste as a significant source of toxic pollution and the emergence of the health effects field of endocrine disruption. Medical-waste incineration is a major source of the dioxin and mercury released into the environment. Almost all humans have measurable residues in their tissues of chlorinated hydrocarbon chemicals, including pesticides, polychlorinated biphenyls (PCBs), and dioxin. In some cases these levels approach the threshold of public health concern.[22] Pressured by advocacy organizations, the health care industry has begun efforts to better manage medical materials and waste. The second development is the emerging toxicological field of endocrine disruption. Endocrine disrupters are a class of chemicals, including many of the persistent organic pollutants, that imitate or block the actions of hormones. These chemicals produce a variety of reproductive and neurodevelopmental disturbances in wildlife, laboratory animals, and humans, often at very low doses. Endocrine disruption, currently the object of renewed study by government, industry, and academia, offers a new toxicological paradigm. Endocrine disruption joins carcinogenesis as the environmental health outcome of concern.[23,24]

Public concern about environmental degradation in both rich and poor nations is developing into a broad environmental health movement.[25] But to argue, as some do, that the quality of human existence is improving because life expectancy is increasing and child mortality is decreasing in many parts of the world is to miss what McMichael has called the "essential newness" of environmental change.[18] The carrying capacity of the earth may appear adequate at this moment in history, particularly for those of us in affluent countries. Economic development and improved access to public health programs have produced expected improvements in less developed countries. Although the world's population has increased fourfold in the last 150 years, the food supply has kept pace. But can we support another approximate doubling of the population by 2050, from 6 billion to 10 or 12 billion, the high-fertility forecast by the United Nations?[26] Will the food supply remain adequate? What are the health consequences of global warming and climate change?[27] What are the consequences of loss of biodiversity, forests, and marine life?[28] Science has only begun to address these questions.[29]

Risk and Caution: Some Definitions

A risk is a potential threat to health and life. How we assess, perceive, communicate, and manage risks is termed *risk analysis*. Risk analysis is a substantive, changing, and controversial field. It is quantitative and produces numerical estimates of exposure, dose, and effect. Data necessary for analysis is often incomplete or absent. Assumptions and estimates must be used. Recently, environmental health scientists using the medical notion of "do no harm" have developed a precautionary principle to guide decision making (see chapter 14).

Patients worried about exposure to environmental or occupational agents often want to know what their risk of developing disease will be. Satisfactory answers are difficult to establish. Risk assessment provides probability figures that apply to populations, not to individuals. And, though derived from quantitative information, the final numbers involve some assumptions and offer only an estimate of how much an exposure to a chemical or physical agent is likely to affect health in a population.[30]

The term *exposure* refers to the intensity and duration of contact with a substance or with a physical agent, such as ultraviolet, x-ray, or microwave radiation. Inhalation, skin absorption, and ingestion are the possible routes of exposure. How much a person actually absorbs from an exposure constitutes the *dose,* the most important determinant in clinical disease. What dose someone was exposed to depends on the amount, route, and duration of exposure and on the concentration and chemistry of the substance or the intensity of the physical agent. The absorbed dose is generally greater if the exposure level is high and, in the case of chemical substances, if the substance involved is nonionized and highly soluble in fat. The biological effect depends on the substance's chemistry or a physical agent's type, on the dose, and on the patient's susceptibility to disease. A brief high exposure, for example, typically produces acute disease.

One of the challenges in assessing an environmental exposure is getting some idea of the actual dose the patient received. The exposure-disease model involves four elements: the exposure to the substance, the dose the patient actually absorbed, the biological effect of the absorbed dose, and the clinical disease that results.[31]

It is important to remember that a final risk figure is a probability, not something guaranteed to happen. Consider, for example, the probability of developing lung cancer from smoking cigarettes—a case in which a clear-cut cause-and-effect relationship exists between exposure and disease. One cannot say unequivocally that any one smoker will develop lung cancer. All that can be said with certainty is that the smoker's risk, or probability, of developing lung cancer is increased by some amount.

Where do risk numbers come from, and how accurate are they? They are derived from a process called *risk assessment,* which has four components:

1. *Hazard identification,* which involves identifying health hazards in the environment and characterizing their physical and chemical properties, their environmental and biological fates, and their abilities to cause disease
2. *Exposure assessment,* which involves estimating human exposure from air, water, food, or skin contact with the hazardous agent
3. *Dose-response assessment,* which involves characterizing the potential of a substance to cause disease as a function of the exposure or dose
4. *Risk characterization,* which integrates information on the agent's toxicity, likely routes of exposure, and dose-response assessment into an estimate of the agent's probable effect on the health of the population

The data used in deriving risk assessments come from clinical studies; from epidemiologic studies of disease in human populations; from animal studies, in which populations of laboratory animals are exposed to large doses of the agent; from in vitro tests of mutagenicity; from their similarity to others that are known; from physiological, biochemical, and toxicological findings; and from comparative metabolic and pharmacokinetic studies. An assessment may weigh the risks of cancer, birth defects, neurotoxicity, immunological changes, or other health effects. To date, most risk assessments have focused on cancer risks, but future risk assessments will likely place more emphasis on immunological and neurological risks and on teratological risks (i.e., risks of fetal damage).

Animal Studies

Clinical and epidemiologic studies offer the most direct means of assessing human health risks associated with hazardous exposures. It is impossible, in most cases, to carry out on humans the kinds of experimental studies needed to make reliable risk predictions. As a result, many epidemiologic studies are done on populations of humans who were exposed to a chemical or a physical agent accidentally or in the course of their work but whose exposure is often difficult to assess. In the case of chronic disease, such as cancer, the long time between an exposure and the appearance of clinical disease makes it difficult to determine the agent's effects on health. In other cases, although a chemical's toxic effects may be apparent, lack of adequate exposure data may make it difficult to use available epidemiologic studies in quantitative risk estimates. Hence, data from laboratory studies usually form the backbone of risk assessment.[32]

Toxicologists readily admit that laboratory animals differ from humans in a variety of ways: humans are larger; live longer; frequently suffer from two or more diseases at a time; may receive different exposures than lab animals do; may metabolize more, store, or excrete the agent differently; and, as a species, are more genetically heterogeneous. Nevertheless, most toxicologist believe, on the basis of experimental evidence, that laboratory animals and humans have more physiological, biochemical, and metabolic similarities than differences.

Extrapolating Results to Humans

One objection to the use of animal data is that lab experiments typically use high doses so that the test will be sensitive and unlikely to give a false negative result; the risk estimate is then extrapolated from results seen at high experimental doses to the much lower levels—often 1,000 to 10,000 times lower—that are typical of environmental exposures. The low-dose estimate is then scaled up to humans, either by a safety-factor approach or by mathematical modeling.

Many risk estimators prefer to use mathematical models to characterize the underlying dose-response curve. In carcinogenesis, risk estimators typically use one of three classes of models. *Tolerance distribution* models assume that each individual in a population has a threshold below which he or she will not respond, and that thresholds vary among individuals. *Mechanistic* models, based on the assumption that cancer originates within a single cell, attempt to characterize the process of transformation mathematically. *Time-to-tumor* models are occasionally used in low-dose extrapolation when data on the time between initial exposure and detection of a tumor in an individual animal are available. The model used can have significant effect on low-dose risk estimates.[32]

Identifying developmental or reproductive toxic agents in humans is possible but is even more complex than identifying carcinogens. The risk assessment must characterize the timing of the exposure, as well as the type and severity of the effects. Reproductive risk assessment is complicated by the fact that a high enough dose of virtually any substance, administered to the mother at the correct time, will have some kind of adverse developmental effect on the fetus.

Refining Risk Assessment

Risk assessment is going through a period of rapid development. Exposure assessment, thought to be one of the most challenging steps in the process, is being refined to consider environmental degradation and accumulation, ingestion of contaminated soil and foliage by cattle, deposition rates of particulates, dermal uptake, biomagnification, pharmacokinetics in human, exposure schedules, and the dietary changes that take place over a lifetime.

New models that try to characterize the exposures of subgroups of a population, children for example, may replace models that emphasize worst-case scenarios (see chapter 15). Risk assessment is also beginning to emphasize, as mentioned above, reproductive and development toxicity, neurotoxicity, and immunotoxicity, as well as carcinogenicity. And finally, new pharmacokinetic models for interpreting rodent and human data may explain differences in biological response between humans and other animals.[31,32]

Closely linked to risk assessment and risk perception is risk communication—the way(s) in which the existence and the relative magnitude of a hazard are communicated. This is too often left to regulators and politicians who know little about health and disease or to health professionals who know little about risk assessment.[33,34]

Physicians and Public Health Professionals

In answering questions about toxic exposure or the other environmental health risks, a physician needs to call on a number skills, including the ability to evaluate and assess the nature and the extent of the patient's exposure to an environmental hazard, the ability to assess the degree to which the hazard presents a health threat, and the ability to communicate the extent of the risk to the patient or to the community so as to minimize or prevent further exposure.[33]

Medical students receive little training in these skills. One study that examined the teaching of occupational and environmental medicine concluded that current medical students receive less than four hours of education in these two areas during their four years of medical school.[35] A decision must be made as to where in the medical school curriculum environmental health and risk assessment should be taught. The most likely candidate is the course on epidmiology (the study of the causes of diseases in populations), but, in general, this subject is given little attention. The same situation applies to the education of physicians after medical school—environmental and occupational health are rarely covered.

An important target of environmental education ought to be physicians. Most physician training is designed around the biomedical model of "finding and fixing" the health problem. We must change the orientation of thinking toward creating health, rather than curing disease. This will require a shift from a bioengineering to an ecological model in medicine. All physicians (especially primary-care physicians) need to understand the relationship of environment to health. They should be able to detect and diagnose environmentally related disease, know how to obtain information about environmental hazards, advise patients on intervention strategies to reduce exposure to environmental hazards, and be able to refer patients to specialists in environmental and occupational medicine.

Most physicians lack an adequate background in statistical reasoning and in environmental and occupational epidemiology, and many risk assessors have not been adequately grounded in human disease biology or in toxicology. There are very few physicians in the U.S. Environmental Protection Agency, and it has be argued that "what is needed is a concerted effort on the part of EPA to hire individuals educated in public health and to provide public health training to present employees."[36]

It is time for physicians and other public health specialists to forge new relationships with environmental specialists and advocates. With the birth of the modern environmental movement in the late 1960s, public health and environmental specialists separated professionally as new departments of environmental protection were created, often out of old public health departments. It is time that these disciplines joined forces to work on the inseparable goals of preserving the environment and promoting the health and well-being of the global population.[4]

Solutions

Given the close connection between environment and health and the fundamental belief systems that prevent us from recognizing the dimensions of the environmental crisis, how can we promote health?

A first step is to stop further environmental degradation. This will require major shifts from policies that control pollution and repair damage to policies that seek to prevent the generation of pollution and environmental damage in the first place.[22]

A second step will be to change the relationship between developed and developing countries. Industrialized nations will need new strategies for transferring technologies, training, and education and for providing financial assistance to non-industrialized nations. These strategies must deal with international debt and must promote economic development that minimizes the destruction of resources and the generation of pollutants while also improving the quality of life.[19] The implementation of such "sustainable development" strategies will require a profound understanding by the developed world of interdependence of all nations when it comes to the global environment.[9]

A third initiative is to develop long-term educational strategies that will change the mindsets of individuals and institutions with respect to protecting the environment and promoting health. Both the citizenry and the leaders of the twenty-first century must understand that virtually every human activity affects the health and welfare of the planet. Air, water, land, and plant and animal life are the bases of human existence. Physicians, engineers, businesspeople, scientists, economists, politicians, and all other professional disciplines must understand environmental issues in relation to health and the quality of life and must bear individual and collective responsibility for stewardship of the world's resources. Shifting our way of thinking to the interdisciplinary, the intergenerational, and the international will demand incorporation of these concepts into the curricula of all disciplines at the undergraduate and graduate levels.[18,25]

To protect the health of populations, we must develop systems of food, energy, and industrial production that can be sustained over generations. We also need value systems of stewardship, precaution, and prevention to guide environmental protection and health promotion. Finding solutions to the threats posed by environmental change is the major health challenge we face.

Acknowledgments

Preparation of this manuscript was supported by a grant from the Jennifer Altman Foundation.

References

1. Hippocrates. Air, water, places: An essay on the influence of climate, water supply and sanitation on health. In: Lloyd GER, editor. *Hippocratic writings*. London: Penguin Books, 1987.

2. McMichael A. Planetary overload: Global environmental change and the health of the human species. Cambridge: Cambridge University Press, 1993.

3. Last JM. Human health in a changing world. In: Public health and human ecology. 2nd ed. Stamford, CT: Appleton and Lange, 1998. pp. 395–425.

4. McCally M, Cassel CK. Medical responsibility and global environmental change. Ann Intern Med 1990; 113:467–473.

5. Leaf A. Potential health effects of global climatic and environmental changes. N Engl J Med 1989; 321:1577–1583.

6. Vitousek PM, Mooney HA, Lubchenco J, Melillo JM. Human domination of earth's ecosystems. Science 1997; 277:494–499.

7. Union of Concerned Scientists. World scientist's warning to humanity. Boston: Union of Concerned Scientists, 1992.

8. Blaustein AR, Wake DB. The puzzling decline of amphibian populations. Sci Amer 1995; 272(4):52–57.

9. United Nations Environment Programme. Global environmental outlook. New York: United Nations Environment Programme, 1997.

10. World Health Organization. Health and environment in sustainable development: Five years after the Earth Summit. Geneva: World Health Organization, 1997.

11. World Resources Institute. World resources: A guide to the global environment 1998–99. Environmental change and human health: The urban environment. Oxford: Oxford University Press, 1997.

12. Clark W. Managing planet earth. Sci Am 1989; 261(3) 45–55.

13. Brown LR, Renner M, Halwell B. Vital Signs 2000. New York: WW Norton, 2000.

14. Daily G. Nature's services: Societal dependence on natural ecosystems. Washington, DC: Island Press, 1997.

15. Ehrlich PR. A world of wounds: Ecologists and the human dilemma. In: Kinne O, editor. Excellence in ecology. vol. 8. Nordbünte, Germany: Ecology Institute, 1993. pp. 184–193.

16. McKibben B. A special moment in history. Atl Mon 1998; 281(5):55–78.

17. Intergovernmental Panel on Climate Change. 1995: Summary for policy makers. Geneva: World Health Organization and United Nations Environment Programme, 1995.

18. McMichael AJ. Global environmental change and human health: New challenges to scientist and policy-maker. J Public Health Policy 1994; 15:407–419.

19. McCally M, Haines A, Fein O, Addington W, Lawrence RS, Cassel CK. Poverty and ill health: Physicians can and should make a difference. Ann Intern Med 1998; 129:726–733.

20. Myers N. Environmental unknowns. Science 1995; 269:368–369.

21. Forman JC, Gardiner BJ, Shinklin J. Large losses of total ozone in Antarctica reveal seasonal ClO_x/NO_x interaction. Nature 1985; 315:207–210.

22. Thornton J, McCally M, Orris P, Weinberg J. Dioxin prevention and medical waste incinerators. Public Health Rep 1996; 11:298–312.

23. Schettler T, Solomon G, Valente M, Huddle A. Generations at risk: Reproductive health and the environment. Cambridge, MA: MIT Press, 1999.

24. Krimsky S. Hormonal chaos: The scientific and social origins of the environmental endocrine hypothesis. Baltimore: Johns Hopkins University Press, 2000.

25. Kotchian S. Perspectives on the place of environmental health and protection in public health and public health agencies. Ann Rev Public Health 1997; 18:245–259.

26. United Nations. World population projections to 2150. New York: United Nations, 1998.

27. McMichael AJ, Haines A, Slooff R, Kovats S, editors. Climate change and human health. Geneva: World Health Organization, 1996.

28. Grifo F, Rosenthal J, editors. Biodiversity and human health. Washington, DC: Island Press, 1997.

29. National Research Council. Our common journey: A transition toward sustainability. Washington, DC: National Academy Press, 1999.

30. Ris CH, Preuss PV. Risk assessment and risk management: A process. In: Gothern CR, Mehlman MA, Marcus WL, editors. Risk assessment and risk management of industrial environmental chemicals. Princeton, NJ: Princeton Scientific, 1988.

31. Gardner MJ. Epidemiological studies of environmental exposure and specific diseases. Arch of Env Health 1990; 43:102–108.

32. Rall DP. Alternatives to using human experience in assessing health risks. Ann Rev Public Health 1987; 8:355–385.

33. National Research Council. Improving risk communication. Washington, DC: National Academy Press, 1989.

34. McCallum DB, Covello VT, Pavlova MT, editors. Effective risk communication: The role and responsibility of government and nongovernment organizations. New York: Plenum Press, 1989. p. 3–16.

35. Levy BS. The teaching of occupational health in U.S. medical schools: Five-year follow-up of an initial survey. Amer J Public Health 1985; 75:79–80.

36. Case BW. Approaching environmental medicine. Pennsylvania Medicine 1992; 93:52–55.

Urban and Transboundary Air Pollution

David C. Christiani and Mark A. Woodin

2

Recognition of the relationship between exposure to air pollutants and respiratory illness dates back to the sixteenth century and vivid descriptions of respiratory disease in metalliferous miners.[1,2] In 1700, the epidemiologist and physician Ramazzini published the first textbook on occupational disease, *De Morbis Artificium Diatriba,* in which he described in detail respiratory diseases due to exposure to silica, cotton, tobacco, flour, and other materials.[3] In fact, much of what we currently understand about environmental lung disease derives from the study of exposed workers since the Industrial Revolution.

Recognition of the relationship between nonworkplace (i.e., community) air pollution and respiratory disease dates back to the first use of coal as a combustion source in the fourteenth century.[4] Later, in the industrial nations of Europe and North America, whole communities were engulfed in air pollutants, resulting in serious illness and death among individuals with cardiopulmonary disease. These episodes occurred in the Meuse Valley of Belgium in 1930; in Donora, Pennsylvania, in 1948; and in London in 1952.[5] These air pollution emergencies were caused by air stagnation, which resulted in greatly increased concentrations of atmospheric pollutants, especially sulfur dioxide and suspended particulates. The worst episode occurred in London, where the number of deaths ultimately reached over 4,000, a total second only to that of the influenza pandemic of 1917–18.[6] The observed marked excess in mortality was due mostly to bronchitis, pneumonia, and acute exacerbation of underlying cardiac and respiratory diseases.

As a result of these epidemics, scientists and governments paid increased attention to the health effects of air pollution—identifying specific pollutant sources and their transport in the atmosphere, elucidating exposure-response relationships, and developing air pollution control technology. Despite these efforts, we face a crisis of worldwide air pollution today. This crisis has several aspects. First, since the atmosphere is dynamic and always changing, contaminants are transported (sometimes over thousands of miles), diluted, precipitated, and transformed. Air pollution, therefore, knows no boundaries or national borders.

Second, the primary emissions of sulfur oxides, nitrogen oxides, carbon monoxide, respirable particulates, and metals (e.g., lead and cadmium) are severely polluting cities and towns in Asia, Africa, Latin America, and Eastern Europe. Many cities in developing nations and in Eastern Europe are experiencing uncontrolled industrial expansion, increasing motor vehicle numbers and congestion, and pollution caused by fuels used for cooking and heating. A 1988 study of developing countries described the relationship between air pollution and wealth (GNP per capita): poorer countries (relying heavily on coal) had significantly higher levels of total suspended particulates (TSPs) than wealthier nations.[7] A more recent report from the World Health Organization (WHO), summarizing a large body of research concerning exposure to various air pollutants, confirms that this disparity between wealthier and poorer nations persists.[8] Some examples follow.

In Beijing, winter days are often dark and gray because of the level of suspended particulates.[9] Indeed, recent research has demonstrated that maximum levels of particulates in Beijing can reach over 1,200 mg/m^3, and mean daily levels in the winter are now routinely over 500 mg/m^3.[8,10,11] The standard for a 24-hour explosure to PM_{10} in the United States is 150 mg/m^3.

In Lanzhou, a city of about 1.3 million located in northwestern China, a four-year mean level of suspended particulates over 1,000 mg/m^3 has been reported.[12] In Shenyang, the capital of Liaoning Province located in northeastern China, a population of over 3 million is subjected to mean winter levels of TSPs in excess of 550 mg/m^3 with maximum levels over 1,100 mg/m^3.[13]

In Delhi, India, mean levels of TSPs also can approach 500 mg/m^3,[14] and in coal-burning townships in South Africa, TSP levels have been found to exceed 1,000 mg/m^3.[15,16] In Nepal, the capital city of Kathmandu suffers from high levels of sulfur dioxide, with mean annual ranges reported in the 270–350 mg/m^3 range.[8]

In the Western Hemisphere, developing areas also struggle with severe air pollution. For example, parts of Mexico City have mean TSP levels of nearly 500 mg/m^3, and maximum levels approach 700 mg/m^3.[17] In Brazil, the city of São Paulo is subjected to high levels of nitrogen dioxide (NO_2), with an annual mean level of 240 mg/m^3 reported.[18]

Third, in nations that have reduced the primary emissions from heavy industry, power plants, and automobiles, new problems have arisen from pollution by newer industries and from air pollution caused by secondary formation of acids and ozone. The resulting high levels of air pollution (including indoor air pollution) contribute heavily to the high mortality rates observed for acute respiratory disease (about 4 million per year) in children under the age of 5 in many developing countries, or approximately 28 percent of all deaths in this age group.[19–22] The danger of air pollution to young children is not limited to developing countries. For example, a report from the Czech Republic showed a doubling of infant mortality from respiratory problems at mean levels of TSP of 68.7 mg/m^3, much lower than is often seen in developing nations.[23]

Finally, although this report is focused on the human health effects of air pollution, there are also many other aspects of the problem. For example, damage to ecosystems and agriculture from acid rain, damage to buildings and artwork, and reduced visibility are all attributable to air pollution. Although industrialization has meant that a large number of pollutants have been released into the environment, in this review, we concentrate on what has been learned about the respiratory health effects of the most common urban air pollutants.

Defining Adverse Health Effects

What constitutes an adverse health effect has been a subject of scientific debate.[24] For the purpose of this review, any effect that results in altered structure or impaired function or that represents the beginning of a sequence of events leading to altered structure or function is considered an "adverse health effect."

Specific Air Pollutants Associated with Adverse Respiratory Effects

Several major types of air pollution are currently recognized to cause adverse respiratory health effects: sulfur oxides and acidic particulate complexes, photochemical oxidants, and a miscellaneous category of pollutants arising from industrial sources. Some of these commonly encountered pollutants are listed in table 2.1.

Table 2.1 Principal air pollutants, their sources, and their respiratory effects

Pollutant	Sources	Health effects
Sulfur oxides, particulates	Coal and oil power plants Oil refineries, smelters Stoves burning wood, coal, kerosene	Bronchoconstriction Chronic bronchitis Chronic obstructive pulmonary disease (COPD)
Carbon monoxide	Motor vehicle emissions Fossil-fuel burning	Asphyxia leading to heart and nervous system damage, death
Oxides of nitrogen (NO_x)	Automobile emissions Fossil-fuel power plants Oil refineries	Airway injury Pulmonary edema Impaired lung defenses
Ozone (O_3)	Automobile emissions Ozone generators Aircraft cabins	Same as NO_x
Polycyclic aromatic hydrocarbons	Diesel exhaust Cigarette smoke Stove smoke	Lung cancer
Radon	Natural	Lung cancer
Asbestos	Asbestos mines and mills Insulation Building materials	Mesothelioma Lung cancer Asbestosis
Arsenic	Copper smelters Cigarette smoke	Lung cancer
Allergens	Pollen Animal dander House dust	Asthma Rhinitis

Source: Adapted, with permission, from H. A. Boushey and D. Sheppard, "Air Pollution," in *Textbook of Respiratory Medicine,* ed. J. F. Murray and J. A. Nadel (Philadelphia: Saunders, 1988).

Sulfur Dioxide and Acidic Aerosols

Sulfur dioxide (SO_2) is produced by the combustion of sulfur contained in fossil fuels, such as coal and crude oil. Therefore, the major sources of environmental pollution with sulfur dioxide are electric power generating plants, oil refineries, and smelters. Some fuels, such as "soft" coal, are particularly sulfur-rich. This has profound implications for nations such as China, which possesses 12 percent of the world's bituminous coal reserves and depends mainly on coal for electric power generation, steam, heating, and (in many regions) household cooking fuel.

Sulfur dioxide is a clear, highly water-soluble gas, so it is effectively absorbed by the mucous membranes of the upper airways, with a much smaller proportion reaching the distal regions of the lung.[25] The sulfur dioxide released into the atmosphere does not remain gaseous. It undergoes chemical reaction with water, metals, and other pollutants to form aerosols. Statutory regulations promulgated in the early 1970s by the U.S. Environmental Protection Agency (EPA) under the Clean Air Act resulted in significant reductions in levels of SO_2 and particulates. However, local reductions in pollution were often achieved by the use of tall stacks, particularly for power plants. This resulted in the pollutants being emitted high into the atmosphere, where prolonged residence time allowed their transformation into acid aerosols. These particulate aerosols vary in composition from area to area, but the most common pollutants resulting from this atmospheric reaction are sulfuric acid, metallic acids, and ammonium sulfates. Sulfur dioxide, therefore, together with other products of fossil-fuel combustion (e.g., soot, fly ash, silicates, metallic oxides) forms the heavy urban pollution that typified old London, many cities in developing nations today that mainly burn coal, and basin regions in the United States, such as in areas of Utah where there are coal-burning plants.

In addition to this smog—a descriptive term generically referring to the visibly cloudy combination of smoke and fog—an acidic aerosol is formed that has been shown to induce asthmatic responses in both adults and children.[26–28] The Harvard Six Cities Study demonstrated a significant association between chronic cough and bronchitis and hydrogen ion concentration—a measure of acidity—rather than sulfate levels or total particulate levels.[29] A more recent report from the Harvard Six Cities researchers involved 13,369 children in 24 communities in Canada and the United States.[30] Two measures of air acidity showed significant effects. Higher particle acidity (nmol/m^3) was significantly associated with an increased risk of bronchitis, while higher levels of gaseous acids were significantly associated with the risk of asthma. The gaseous acids were also positively associated with wheeze attacks, chronic wheezing, and any asthmatic symptoms. A time-series analysis that examined daily mortality and morbidity in Buffalo, New York, demonstrated significant positive associations between aerosol acidity and increased respiratory hospital admissions and respiratory mortality.[31]

Both epidemiologic and controlled human studies have demonstrated remarkable sensitivity of persons with asthma while exercising to the bronchoconstrictive effects of acidic aerosols.[32,33] Several studies have also linked exposure to acidic aerosols and mortality, documenting an increase in deaths of persons with underlying chronic heart and lung disease who had been exposed.[34,35] Finally, acidic aerosols result in "acid rain," which may threaten aquatic life.[36]

The acute bronchoconstrictive effects of sulfur dioxide have been well known for almost 30 years.[37,38] For example, inhalation of high concentrations of SO_2 has been shown to increase airway resistance in healthy normal volunteers[37,38] and profoundly in asthmatics.[32,33,39] Although this effect in healthy humans is noted only after inhalation of high concentrations (e.g., 5 ppm), well above the concentrations found in most polluted cities, other human studies have shown that patients with asthma are exquisitely sensitive to the bronchoconstrictive effects of SO_2 and react to much lower levels.[39–42] Because SO_2 is highly water soluble, nearly all of the inspired SO_2 gas is removed in the upper airways during rest. Exertion (work or exercising) will increase the fraction of SO_2 gas to the lower airways and therefore help to precipitate bronchoconstriction. It has been demonstrated that concentrations of SO_2 as low as 0.4 ppm can cause symptomatic bronchoconstriction when inhaled by exercising asthmatics. These levels may be encountered in polluted urban air, and exposure to these levels for even several minutes is enough to induce bronchospasm.[43] Thus, sulfur-dioxide-induced bronchoconstriction is not necessarily prevented by the current EPA standard of 0.14 ppm as a 24-hour maximal average, because this standard does not set limits for maximal concentrations over shorter periods (table 2.2).

Table 2.2 U.S. national air quality standards for six criteria pollutants

Pollutant	Maximum allowable exposure	Time period
Carbon monoxide	9 ppm	8-hour average
	35 ppm	1-hour average
Ozone	0.12 ppm	1-hour average
	0.08 ppm*	8-hour average
PM_{10}	50 $\mu g/m^3$	Annual arithmetic mean
	150 $\mu g/m^3$	24-hour average
$PM_{2.5}$†	15 $\mu g/m^3$	Annual arithmetic mean
	65 $\mu g/m^3$	24-hour average
Lead	1.5 $\mu g/m^3$	Quarterly average
Sulfur dioxide	0.03 ppm	Annual arithmetic mean
	0.14 ppm	24-hour average
	0.50 ppm	3-hour average
Nitrogen dioxide	0.053 ppm	Annual arithmetic mean

*Proposed in 1997 by EPA to replace the 1-hour standard. Currently under judicial review.

†Proposed in 1997 by EPA to supplement PM_{10} standard. Currently under judicial review.

In addition to the acute bronchoconstrictive effects of sulfur dioxide, there is epidemiologic evidence for chronic airway obstruction in persons exposed to elevated levels of SO_2.[44,45]

Particulates

Particulate air pollution is closely related to SO_2 and aerosols. The term usually refers to particles suspended in the air after various forms of combustion or other industrial activity. In the epidemics noted earlier, the air pollution was characterized by high levels of particulates, sulfur dioxide, and moisture. Recent studies have shown that particulate air pollution per se was associated with increases in daily mortality in London in both the heavy smog episodes of the 1950s and the lower pollution levels of the late 1960s and early 1970s.[46,47]

A study performed in Steubenville, Ohio, between 1974 and 1984 revealed that total suspended particulate counts were significantly associated with increased daily mortality, with season and temperature controlled for.[48] An increase in particulates of 100 $\mu g/m^3$ was associated with a

4 percent increase in mortality on the succeeding day. The striking quantitative agreement in the relative increase in mortality among studies done in London, New York, and Steubenville further strengthened the evidence for a causal association. More recently, Samet and colleagues examined the effect of air pollution by particles <10 microns in diameter (PM_{10}) in 20 U.S. cities from 1987 to 1994.[49] The cities involved covered the entire United States, increasing the study's generalizability. The major finding for PM_{10} was that an increase of 10 $\mu g/m^3$ corresponded to a significant increase in mortality of 0.68 percent.

Recent interest in particulate air pollution has focused in particle size. Specifically, particles <2.5 microns in diameter ($PM_{2.5}$, also called "fine" particles), produced almost exclusively from combustion sources, have been closely studied under the hypothesis that such small particles can penetrate deeply into the lung, while larger particles would be trapped in the upper airway. A detailed study by Harvard Six Cities researchers used tracer metals to split fine particles into those from mobile sources (e.g., motor vehicle exhaust), coal combustion, and crustal particles (e.g., dirt).[50] Both mobile sources and coal combustion $PM_{2.5}$ were significantly associated with daily mortality. A 10 $\mu g/m^3$ increase in fine particles from mobile sources was associated with a 3.4 percent increase in daily mortality, while the same increase in fine particles from coal combustion was associated with a 1.1 percent increase in daily mortality. Interestingly, the fine particles contributed by crustal sources had little effect on daily mortality.

In a reanalysis of three recent longitudinal studies on elementary school children, Schwartz and Neas found that increases in fine particles were significantly associated with respiratory symptoms and decrements in peak flow measurements.[51] A noteworthy finding was that coarser particles in the $PM_{2.5}$–PM_{10} range were not significantly associated with the health outcomes of interest.

Research has begun to examine the relationship of particulate levels with subtle changes in heart-rate variability (HRV), a measure of the cyclic variations of beat-to-beat intervals and of the instantaneous heart rate.[52,53] A complete discussion of the complexities of HRV is beyond the scope of this chapter, but excellent summaries are available.[52] The clinical implications of changes in HRV are also not completely understood; however, there is evidence that reduced HRV is a risk factor for adverse cardiac events, including angina, myocardial infarction (MI), life-threatening arrhythmias, and congestive heart failure.[52,53] The Framingham Heart Study found that in a prospective evaluation of subjects with no suspected heart disease, reduced HRV was significantly associated with a nearly 50 percent increase in new adverse cardiac events.[53]

The effect of increased levels of fine particles on HRV was examined by Gold and colleagues.[54] Using continuous monitoring of heart rate for 25 minutes per week, including five minutes each of standing, outdoor exercise, recovery from exercise, and rest, the researchers demonstrated a significant reduction in HRV associated with higher levels of fine-particle air pollution. The effect was strongest when ozone and 24-hour fine-particle levels were used in the statistical model. Another study done in the Utah Valley in central Utah found that all measures of HRV were significantly reduced by increased levels of PM_{10}.[55]

A current limitation of the research on air pollution and HRV is the almost exclusive use of older subjects, often with comorbid health conditions. Whether particulate air pollution has the same effects on the HRV of younger individuals is not known. The fact that many epidemiologic studies have demonstrated increased cardiovascular morbidity and mortality associated with higher levels of particulate air pollution clearly makes the clarification of the role of HRV an urgent research priority.

The very high particulate levels (both indoor and outdoor) measured in developing countries (table 2.3) range up to 100 times the current U.S. standard of 150 $\mu g/m^3$ for PM_{10}.[56] The health consequences, as one would expect from the above discussion, are great. Deaths in children from acute respiratory disease are high when particulate levels are high, even in nations with lower overall infant mortality.[19] For example, Western Europe has a low overall infant mortality rate but has three times North America's rate of infant respiratory-disease mortality.[19] The rate in Eastern Europe is nearly 20 times that in North America. After infant diarrhea, acute respiratory disease is believed to be the major cause of death in children under the age of five in the developing world and to account for up to 4 million fatalities per year.[19] Air pollution, both indoor and outdoor, appears to play a central role in this epidemic.

Photochemical Oxidants

The other two most commonly generated industrial and urban pollutants are ozone and oxides of nitrogen. In contrast to sulfur dioxide, these two

Table 2.3 Indoor air pollution in developing countries

Country	Conditions	Particulate concentration (μg/m^3)
New Guinea	Overnight	200–9,000
Kenya	Highlands	2,700–7,900
	Lowlands	300–1,500
	24 hours	1,200–1,900
India	Cooking material	
	Wood	15,000
	Dung	18,300
	Charcoal	5,500
	General	4,000–21,000
Nepal	Cooking (wood)	4,700
China	Cooking (wood)	2,600
Gambia	24 hours	1,000–2,500
During cooking		
India	Villages	8,200
Nepal	Villages*	8,200
	Villages†	3,000

Source: Reproduced, with permission, from reference 56.
*Before introduction of new cooking stoves.
†After introduction of new cooking stoves.

substances are produced not so much by heavy industry as by the action of sunlight on the waste products of the internal combustion engine.[57] The most important of these products are unburned hydrocarbons and nitrogen dioxide (NO_2), a product of the interaction of atmospheric nitrogen with oxygen during high-temperature combustion. Ultraviolet irradiation by sunlight of this mixture in the lower atmosphere results in a series of chemical reactions that produce ozone, nitrates, alcohols, ethers, acids, and other compounds that appear in both gaseous and particulate aerosols.[58] This mixture of pollutants constitutes the smog that has become associated with cities with ample sunlight (Los Angeles, Mexico City, Taipei, Bangkok) and with most other urban areas in moderate climates.

Ozone and nitrogen dioxide have been studied extensively in both animals and humans. Both gases are relatively water insoluble and reach lower into the respiratory tract. Therefore, these gases can cause damage at any site from the upper airways to the alveoli.[59]

Exposure to Ozone and Nitrogen Dioxide

Clinical research on the acute effects of exposure to ozone reveals symptoms and changes in respiratory mechanics at rest or during light exercise after exposure to 0.3–0.35 ppm.[60–62] In exercising subjects, similar effects will occur with exposure to 150 ppb, a level currently found in urban air.[63] There is also recent evidence that the upper-airway (nose) inflammation induced by ozone correlates with the lower-airway response, as demonstrated by simultaneous nasal and bronchoalveolar lavage.[64]

Animal studies have clearly documented severe damage to the lower respiratory tract after high exposure, even briefly, to ozone. The pulmonary edema noted is similar to that noted in humans accidentally exposed to lethal levels.[65] Animals exposed to sublethal doses develop damage to the trachea, the bronchi, and the acinar region of the lung.[66]

Levels of ozone (O_3) that exceed the federal standard (120 ppb for 1 hour) are frequent in cities in Southern California and in the Northeast. Large air masses with ozone concentrations that exceed the federal standard have been observed under stable meteorological conditions that can last several days in the Northeast. High concentrations of O_3 are most often observed in the summertime, when sunlight is most intense and temperatures are highest—conditions that increase the rate of photochemical formation. Strong diurnal variations also occur, with O_3 levels generally lowest in the morning hours, accumulating through midday, and decreasing rapidly after sunset.

Epidemiologic studies have demonstrated significant adverse respiratory effects associated with increased ozone even at levels well below 120 ppb. Gold and colleagues examined the effect of ozone levels and particulates in Mexico City on peak expiratory flow in 40 schoolchildren ages 8–11.[67] In the study, mean daily ozone levels were 52 ppb, less than half of the 120 ppb 1-hour standard. The researchers found that both particulates and ozone levels contributed to an observed significant decrement in morning peak flow. The effects of ozone were evident more quickly (0–4 days) than those of particulates (4–7 day lag period). The combined effect of a one-week exposure to the interquartile range of 25 ppb ozone and 17 $\mu g/m^3$ $PM_{2.5}$ was a statistically significant 7.1 percent decline in morning peak expiratory flow.

Kinney and Lippmann examined 72 cadets from the U.S. Military Academy.[68] For summer training, 21 cadets went to Fort Dix, New Jersey, an area associated with high ozone. The remaining cadets attending summer training in other areas associated with more moderate levels of ozone (e.g., Fort Leonard Wood, Missouri). The cadets training at Fort Dix (mean 1-hour ozone level of 71.3 ppb) had a mean decline in lung function as measured by forced expiratory volume in 1 second (FEV_1) more than double that exhibited by the cadets who trained at other centers (mean ozone levels of 55.6, 55.4, and 61.7 ppb for the three centers).

Of interest in the Kinney and Lippmann study is the fact that mean ozone levels at all centers were well below the 120-ppb level. (Indeed, levels were closer to each other than to 120 ppb.) This relatively low level of ozone pollution was also evident in the Gold et al. study cited above. One effect of studies showing adverse health effects of ozone at levels well below the 120-ppb standard was a changing of the ozone standards by EPA in 1997. The 1-hour limit was eliminated and replaced with an 8-hour, 80 ppb standard. Unfortunately, the bulk of epidemiologic studies, including those cited above, suggest that even this standard may not be conservative enough.

While O_3 concentrations may be elevated in outdoor air, they are substantially lower indoors. The lower indoor concentrations are attributed to the reaction of O_3 with various surfaces, such as walls and furniture. Mechanical ventilation systems also remove O_3 during air-conditioning. For these reasons, staying indoors or closing car windows and using air-conditioning are generally recommended as protecting against exposure to ambient O_3.

Studies of the effect of nitrogen dioxide on healthy human subjects have demonstrated results similar to those found with ozone, with the exception that the concentration of nitrogen dioxide shown to produce mechanical dysfunction is above the concentrations noted in pollution episodes (i.e., 2.5 ppm).[69] In animal studies, similar damage to alveolar cells is noted after exposure to relatively high concentrations of nitrogen dioxide.[66]

In any case, current epidemiologic data show that exposure to photochemical oxidants, particularly ozone, can cause bronchoconstriction in both normal and asthmatic people.[70,71]

Carbon Monoxide (CO)

Carbon monoxide is emitted mainly from internal combustion engines used in motor vehicles. In the United States, emission controls for new vehicles resulted in a 77 percent reduction in CO emissions between 1975 and 1981. This decline alone has been responsible for substantial improvement in outdoor CO levels.[72] For example, in the Los Angeles and Orange County basins of California, third-quarter daily maximum-hour CO concentrations dropped from approximately 10 ppm in 1970 to 3 ppm in the 1980s.[73] However, because of increased vehicle use and traffic congestion, many localized "hot spots" in U.S. cities remain where CO levels still exceed air-quality standards. Exposure can also occur after the burning of coal, paper, wood, oil, gas, or any other carbonaceous material.

The health effects of carbon monoxide have been documented in clinical observations of patients with CO intoxication and in experimental and epidemiologic studies of persons exposed to low-level CO. Carbon monoxide is an odorless, colorless gas produced by incomplete combustion of carbonaceous fuels such as wood, gasoline, and natural gas.[74] Because of its marked affinity for hemoglobin, CO impairs oxygen transport, and poisoning often manifests as adverse effects in the cardiovascular and central nervous systems, with the severity of the poisoning directly proportional to the level and duration of the exposure. Thousands of people die annually (at work and at home) from CO poisoning, and an even larger number suffer permanent damage to the central nervous system. Sizable portions of the workers in any country have significant CO exposure, as do a larger proportion of persons living in poorly ventilated homes where biofuels are burned.

The pathophysiology of CO poisoning can be conceptualized as antioxygen activity, since CO is an antimetabolite of oxygen. Inhaled CO binds strongly to hemoglobin in the pulmonary capillaries, resulting in the complex called carboxyhemoglobin (COHb). The affinity of human hemoglobin for CO is about 240 times its affinity for oxygen.[75] The formation of COHb has two considerable effects: it blocks oxygen carriage by inactivating hemoglobin, and its presence in the blood shifts the dissociation curve of oxyhemoglobin to the left so that the release of remaining oxygen to tissues is impaired. Because of this latter effect, the presence of any level of COHb in the blood, above the background from catabolism of 0.3–0.8 percent, will interfere with tissue oxygenation much more than an equivalent reduction in hemoglobin from anemia or bleeding. Carbon monoxide also binds with myoglobin to form carboxymyoglobin, which may disturb muscle metabolism (especially in the heart).

The clinical effects of acute CO poisoning vary with the level of COHb, ranging from nonspecific symptoms (headache, dizziness, fatigue) to death.[76] Generally, acute symptoms appear at COHb levels of 10 percent or greater, and levels above 50 percent are associated with collapse, convulsions, and death. Because of the profound physiological effects associated with increasing levels of COHb, individuals with certain underlying conditions are at particularly high risk for CO poisoning. These conditions include chronic hypoxemia from lung or heart disease, cerebrovascular disease, peripheral vascular disease, arteriosclerotic heart disease, anemia, and hemoglobinopathies.[76] Fetuses and children are more susceptible than adults to CO poisoning.[77,78]

Chronic, lower-level exposure to CO has also been postulated to accelerate atherosclerotic vascular disease by affecting cholesterol uptake in the arterial wall, though results from animal and in vitro studies are conflicting.[79–82] There is also some evidence that CO may accelerate clot lysis time[83] and may increase platelet activity and coagulation.[84] These alterations in the clotting system could increase the risk for thromboembolism in the heart or the brain. Such changes would be consistent with the epidemiologic findings that higher levels of CO are significantly associated with increased cardiovascular hospital admissions in Tucson, Arizona,[85] and in eight urban U.S. counties.[86]

In addition to containing sulfur dioxide, particulates, and photochemical pollutants, urban air contains a number of known carcinogens, including polycyclic aromatic hydrocarbons (PAHs), n-nitroso compounds, and, in many regions, arsenic and asbestos. Exposure to these compounds is associated with increased risk of lung cancer in various occupationally exposed groups (e.g., coke-oven workers exposed to PAHs and insulators exposed to asbestos). Therefore, populations living near coke ovens or exposed to asbestos insulation at home and in public buildings also may be at increased risk of lung cancer.

In addition to these agents, airborne exposure to the products of waste incineration, such as dioxins and furans, may be on the increase in some communities. Though usually present in communities in concentrations much lower than those found in workplaces, airborne dioxins, furans, and other incineration products may still lead to increased lung cancer risks, particularly in neighborhoods or villages near point sources where their levels may be substantial. The degree of cancer risk for ambient exposures to these compounds has not been calculated to date.

Indoor Air Pollution

No description of the health effects of air pollution is complete without a discussion of indoor air pollution. The descriptions above generally have referred to community ("ambient") air pollution. Moreover, as mentioned earlier, much of what we now know about the health effects of airborne toxins derives from the study of workers in factories and mines—both "indoor" environments. However, today, both the common and the scientific use of the term *indoor air pollution* refer to homes and non-factory public buildings (e.g., modern office buildings, hospitals). Most indoor environments, whether traditional village homes in developing countries or tightly sealed high-rise office buildings, have air pollutant sources. Pollution can come from heating and cooking combustion, pesticides, tobacco emissions, abrasion of surfaces, evaporation of vapors and gases, radon, and microbiological material from people and animals.[87] High concentrations of pollution in indoor settings in either the developed or developing world can be associated with mucous-membrane irritation, discomfort, illness, and even death.[88]

Although indoor air pollution has increased substantially in the industrialized nations because of tighter building construction and because of the widespread use of building materials that may give off gaseous chemicals, indoor air pollution is a particular problem in the poor communities of developing countries. Indeed, it is estimated by WHO that 2 to 3 million deaths annually in developing nations are caused by exposure to severe indoor air pollutants and that a disproportionate number of these deaths are in women and young children.[8,20] Wood, crop residues, animal dung, and other forms of biomass are used by approximately half the world's population (about 3 billion people) but by 90 percent of people in rural areas of developing nations as cooking and/or heating fuels, often in poorly ventilated conditions.[20] For example, in sub-Saharan Africa, India, China, and Southeast Asia and its associated island nations, approximately

80 percent, 75 percent, 85 percent, and 65 percent, respectively, of the people use coal or biomass as household fuel.[89] This leads to high exposures to such air pollutants as carbon monoxide, particulates, and polycyclic aromatic hydrocarbons, particularly for women and children. A WHO summary of studies on indoor air pollution in developing countries shows levels of PM_{10} in the range of 300–3,000 $\mu g/m^3$, reaching as high as 30,000 $\mu g/m^3$ during cooking.[20] Homes using biomass for cooking and heating in developing nations have mean 24-hour CO levels of 2–50 ppm, rising to 10–500 ppm during cooking.[20] For reference, the U.S. EPA's current standard for mean 24-hour PM_{10} is 150 $\mu g/m^3$ and for CO the 24-hour maximum is 9 ppm. The health effects of the high indoor air pollution exposures experienced by people in developing countries are summarized in table 2.1. It is clear, however, that the extremely high exposures to particulate and CO must contribute to the fact that, in less developed countries, 9.4 percent of lost disability adjusted life years (DALYs) are explained by acute respiratory infections (mostly in young children).[89] This percentage is only 1.6 percent for the world's more developed nations.[89]

There are clearly strong indications that indoor air pollution is associated with acute respiratory infection in infants and children under age five. In view of the large numbers of children from developing countries who die from respiratory infection each year, this is a high priority for public health research and action.[88]

Current Regulation of Air Pollution

Approaches to controlling air pollution have varied from nation to nation. For example, in Great Britain, the approach was to recognize that outdoor urban air—not specified as to pollutants—was unhealthy, and the government limited emission of a mixture of suspected pollutants with its Clean Air Act of 1956.

The approach in the United States has been different. Congress passed a Clean Air Act in 1970, which required the EPA to define air-quality standards "allowing a margin of safety" to protect the public's health. This required identifying specific pollutants and setting standards for each. The pollutants chosen as criteria were sulfur dioxide, carbon monoxide, nitrogen dioxide, ozone, lead, and total suspended particulates. Amendments to the Clean Air Act passed in 1990 replaced TSP with PM_{10}, a more relevant biological measure.

The assumption underlying this entire approach is that scientific research would allow the identification of a threshold concentration for each pollutant, below which adverse health effects would not occur. Even with the inherent problems of this assumption, standards were set for all six pollutants (see table 2.2). Therefore, the federal government prescribed procedures and standards for the states to follow. The laws dictate that polluters file permits and maintain documentation and that failure to comply will result in potential fines, loss of operating permits, liability for cleanup, and civil or even criminal penalties for cleanup and/or injury. There are obvious instances where compliance with the law still does not ensure health protection against air pollutants, or even environmental protection. For example, in the construction of new highways, the environmental-impact review process assesses compliance with air quality standards at outdoor locations. Ignored in the assessment for the highways are other important impacts, such as the associated industrial and construction projects, the increase in miles traveled in vehicles (and the consequent emissions), and the increased general energy expenditure for utilities such as electricity.

The limitations of the current U.S. approach to air pollution can be briefly summarized as follows:

Ozone and lead are clearly not adequately controlled under the current standards (Section 109 of the Clean Air Act). Also, there is inadequate control of acidity inherent in the SO_2 limits. Moreover, in Section 112 (which addresses carcinogens), the U.S. government's approach has been not to force new technological development of controls, but rather to use the least expensive available methods.

There has been a failure to address indoor air pollution.

There has been a failure to adequately address noncriteria airborne pollutants that affect human health through indirect mechanisms—that is, CO_2 and chlorofluorocarbon emissions as causes of global warming and stratospheric ozone depletion, respectively.

Even the new limits set on ozone, PM_{10}, and $PM_{2.5}$ in 1997, assuming they survive their current court challenges, appear inadequate to protect public health.

The problem of controlling air pollution is indeed a pressing one. Atmospheric pollution has now reached a level that threatens not only the health of entire populations, but also their survival. Various national regulatory approaches have not, so far, been up to the task of controlling pollution on a global scale. Air pollution is a growing, global problem. Only global approaches will succeed in controlling it.

References

1. Agricola G. *De re metallica.* Basel; 1556. Translation in *Mining Magazine* (London), 1912.

2. Paracelsus T. *Von der Bergsacht.* Dilingen, 1567.

3. Ramazzini B. *De morbis artificium diatriba.* Translation: *Disease of workers.* Chicago: University of Chicago Press, 1940.

4. Ayres SM, Evans RG, Buehler ME. Air pollution: A major public health problem. *CRC Crit Rev Clin Lab Sci* 1972; 3:1–40.

5. Ad Hoc Committee on the Health Effects of Air Pollution, American Thoracic Society. *Health effects of air pollution.* New York: American Thoracic Society, Medical Section of American Lung Association, 1978.

6. Logan WPD. Mortality in the London fog incident. *Lancet* 1953; 1:336–339.

7. Smith KR. Air pollution: Assessing total exposure in developing countries. *Environ* 1988; 30:16.

8. *Guidelines for air quality.* Geneva: World Health Organization, 2000.

9. Urban air pollution in the People's Republic of China, 1981–1984. Geneva: World Health Organization, 1985. Report No: EHE-EFP/85–5.

10. Wang X, Ding H, Ryan L, Xu X. Association between air pollution and low birth weight: A community-based study. *Environ Health Perspect* 1997; 105:514–520.

11. Xu X, Dockery DW, Christiani DC, Li B, Huang H. Association of air pollution with hospital outpatient visits in Beijing. *Arch Environ Health* 1995; 50:214–220.

12. Qian Z, Chapman RS, Tian Q, Chen Y, Lioy PJ, Zhang J. Effects of air pollution on children's respiratory health in three Chinese cities. *Arch Environ Health* 2000; 55:126–133.

13. Yu D, Jing L, Xu X. Air pollution and daily mortality in Shenyang, China. *Arch Environ Health* 2000; 55:115–20.

14. Chhabra SK, Chhabra P, Rajpal S, Gupta RK. Ambient air pollution and chronic respiratory morbidity in Delhi. *Arch EnvironHealth* 2001; 56:58–64.

15. Kemeny E, Ellerbeck RH, Briggs AB. An assessment of smoke pollution in Soweto: Residential air pollution. In: *Proceedings of the National Association for Clean Air,* 1988 Nov 10–11; Pretoria, South Africa.

16. Tshangwe HT, Kgamphe JS, Annegarn HJ. Indoor-outdoor air particulate study in a Soweto home: Residential air pollution. In: *Proceedings of the National Association for Clean Air,* 1988 Nov 10–11; Pretoria, South Africa.

17. Romieu I, Meneses F, Sienra-Monge JJ, Huerta J, Ruiz Velasco S, et al. Effects of urban air pollutants on emergency visits for childhood asthma in Mexico City. *Am J Epidemiol* 1995; 141:546–553.

18. Saldiva PH, Pope CA III, Schwartz J, Dockery DW, Lichtenfels AJ, Salge JM, et al. Air pollution and mortality in elderly people: A time series study in São Paulo, Brazil. *Arch Environ Health* 1995; 50:159–164.

19. Leowski J. Mortality from acute respiratory infections in children under 5 years of age: Global estimates. *World Health Stat Q* 1986; 39:138–144.

20. Bruce N, Perez-Padilla R, Albalak R. Indoor air pollution in developing countries: A major environmental and public health challenge. *Bull World Health Organ* 2000; 78:1078–1092.

21. Acute respiratory infections: The forgotten pandemic. *Bull World Health Organ* 1998; 76:101–103.

22. Campbell H. Acute respiratory infections: A global challenge. *Arch Dis Child* 1995; 73: 281–283.

23. Bobak M, Leon DA. The effect of air pollution on infant mortality appears specific for respiratory causes in the postneonatal period. *Epidemiology* 1999; 10:666–670.

24. Guidelines as to what constitutes an adverse respiratory health effect, with special reference to epidemiologic studies of air pollution. *Am Rev Respir Dis* 1985; 131:666–668.

25. Frank NR, Yoder RE, Brain JD, Yokoyama E. SO_2 (^{35}S-labeled) absorption by the nose and mouth under conditions of varying concentrations and flow. *Arch Environ Health* 1969; 18:315–322.

26. Bates DV, Sizto R. Air pollution and hospital admissions in Southern Ontario: The acid summer haze effect. *Environ Res* 1987; 43:317–331.

27. Bates DV, Sizto R. The Ontario air pollution study: Identification of the causative agent. *Environ Health Perspect* 1989; 79:69–72.

28. Ostro BD, Lipsett MJ, Wiener MB, Selner JC. Asthmatic responses to airborne acidic aerosols. *Am J Public Health* 1991; 81:694–702.

29. Ware JH, Ferris BG Jr, Dockery DW, Spengler JD, Stram DO, Speizer FE. Effects of ambient sulfur dioxides and suspended particles on respiratory health of preadolescent children. *Am Rev Respir Dis* 1986; 133:834–842.

30. Dockery DW, Cunningham J, Damokosh AI, Neas LM, Spengler JD, Koutrakis P, et al. Health effects of acid aerosols on North American children: Respiratory symptoms. *Environ Health Perspect* 1996; 104:500–505.

31. Gwynn RC, Burnett RT, Thurston GD. A time-series analysis of acidic particulate matter and daily mortality in the Buffalo, New York region. *Environ Health Perspect* 2000; 108:125–133.

32. Bauer MA, Utell MJ, Morrow PE, Speers DM, Gibb FR. Inhalation of 0.30 ppm nitrogen dioxide potentiates exercise-induced bronchiospasm in asthmatics. *Am Rev Respir Dis* 1986; 134:1203–1208.

33. Koenig JQ, Covert DS, Pierson WE. Effects of inhalation of acidic compounds on pulmonary function in allergic adolescent subjects. *Environ Health Perspect* 1989; 79:173–178.

34. Thurston GD, Ito K, Lippmann M, Hayes C. Reexamination of London, England, mortality in relation to exposure to acidic aerosols during 1963–1971 winters. *Environ Health Perspect* 1989; 79:73–82.

35. Ostro BD. A search for a threshold in the relationship of air pollution to mortality: A reanalysis of data on London winters. *Environ Health Perspect* 1984; 58:397–399.

36. German E, Martin F, Litzau JT. Acid rain: Ionic correlations in the Eastern United States, 1980–1981. *Science* 1984; 225:407–409.

37. Frank NR, Amdur MO, Worcester J, Whittenberger JL. Effects of acute controlled exposure to SO_2 on respiratory mechanics in healthy male adults. *J Appl Physiol* 1962; 17:252–258.

38. Peden DB. Mechanisms of pollution—Induced airway disease: In vivo studies. *Allergy* 1997; 52:37–44.

39. Sheppard D, Wong WS, Uehara CF, Nadel JA, Boushey HA. Lower threshold and greater bronchomotor responsiveness of asthmatic subjects to sulfur dioxide. *Am Rev Respir Dis* 1980; 122:873–878.

40. Sheppard D, Saisho A, Nadel JA. Exercise increases sulfur dioxide-induced bronchoconstriction in asthmatic subjects. *Am Rev Respir Dis* 1981; 123:486–491.

41. Koenig JQ, Pierson WE, Horike M, Frank R. Effects of SO_2 plus NaCl aerosol combined with moderate exercise on pulmonary function in asthmatic adolescents. *Environ Res* 1981; 25:340–348.

42. Linn WS, Venet TG, Shamoo DA, Valencia LM, Anzar UT, Spier CE, et al. Respiratory effects of sulfur dioxide in heavily exercising asthmatics: A dose-response study. *Am Rev Respir Dis* 1983; 127:278–283.

43. Sheppard D, Eschenbacher WL, Boushey HA, Bethel RA. Magnitude of the interaction between the bronchomotor effects of sulfur dioxide and those of dry (cold) air. *Am Rev Respir Dis* 1984; 130:52–55.

44. Xu XP, Dockery DW, Wang LH. Effects of air pollution on adult pulmonary function. *Arch Environ Health* 1991; 46:198–206.

45. Holland WW, Reid, DD. The urban factor in chronic bronchitis. *Lancet* 1965; 1:445–448.

46. Mazumdar S, Schimmel H, Higgins ITT. Relation of daily mortality to air pollution: An analysis of 14 London, England winters, 1958/59–1971/2. *Arch Environ Health* 1982; 37:213–220.

47. Schwartz J, Marcus A. Mortality and air pollution in London: A time-series analysis. *Am J Epidemiol* 1990; 131:185–194.

48. Schwartz J, Dockery DW. Particulate air pollution and daily mortality in Steubenville, Ohio. *Am J Epidemiol* 1992; 135:12–19.

49. Samet JM, Dominici F, Curriero FC, Coursac I, Zeger SL. Fine particulate air pollution and mortality in 20 U.S. cities, 1987–1994. *N Engl J Med* 2000; 343:1742–1749.

50. Laden F, Neas LM, Dockery DW, Schwartz J. Association of fine particulate matter from different sources with daily mortality in six U.S. cities. *Environ Health Perspect* 2000; 108:941–947.

51. Schwartz J, Neas LM. Fine particles are more strongly associated than coarse particles with acute respiratory health effects in schoolchildren. *Epidemiology* 2000; 11:6–10.

52. Task Force of the European Society of Cardiology and the North American Society of Pacing and Electrophysiology. Heart rate variability: Standards of measurement, physiologic interpretation, and clinical use. *Circulation* 1996; 93:1043–1065.

53. Tsuji H, Larson MG, Venditti FJ Jr, Manders ES, Evans JC, Feldman CL, et al. Impact of reduced heart rate variability on risk for cardiac events. Circulation 1996; 94:2850–2855.

54. Gold DR, Litonjua A, Schwartz J, Lovett E, Larson A, Nearing B, et al. Ambient pollution and heart rate variability. *Circulation* 2000; 101:1267–1273.

55. Pope CA III, Verrier RL, Lovett EG, Larson AC, Raizenne ME, Kanner RE, et al. Heart rate variability associated with particulate air pollution. *Am Heart J* 199; 138:890–899.

56. Pandey MR, Boleij JS, Smith KR, Wafula EM. Indoor air pollution in developing countries and acute respiratory infection in children. *Lancet* 1989; 1:427–429.

57. *Photochemical oxidants—Environmental health criteria* 7. World Health Organization Report CIS 80–1919. Geneva: World Health Organization, 1979.

58. Niki H, Dahy EE, Weinstock B. Mechanisms of smog reactions. *Adv Chem* 1972; 113:16–57.

59. Miller JF. Similarity between man and laboratory animals in regional pulmonary deposition of ozone. *Environ Res* 1978; 17:84–101.

60. Bates DV, Bell GM, Burnham CD, Hazucha M, Mantha J, Pengelly LD, et al. Short-term effects of ozone on the lung. *J Appl Physiol* 1972; 32:176–181.

61. Hazucha M, Silverman F, Parent C, Field S, Bates DV. Pulmonary function in man after short-term exposure to ozone. *Arch Environ Health* 1973; 27:183–188.

62. Hackney JD, Linn WS, Buckley RD, Pedersen EE, Karuza SK, Law DC, et al. Experimental studies on human health effects of air pollutants: I. Design considerations. *Arch Environ Health* 1975; 30:373–378.

63. Hackney JD, Linn WS, Mohler JG, Collier CR. Adaptation to short-term respiratory effects of ozone in men repeatedly exposed. *J Appl Physiol* 1977; 43:82–85.

64. Graham DE, Koren HS. Biomarkers of inflammation in ozone-exposed subjects. *Am Rev Respir Dis* 1990; 142:152–156.

65. Cheng PW, Boat TF, Shaikh S, Wang OL, Hu PC, Costa DL. Differential effects of ozone on lung epithelial lining, fluid volume, and protein content. *Exp Lung Res* 1995; 21:351–365.

66. Dungworth D, Goldstein E, Ricci PF. Photochemical air pollution—Part II. *West J Med* 1985; 142:523–531.

67. Gold DR, Damokosh AI, Pope CA III, Dockery DW, McDonnell WF, Serrano P, et al. Particulate and ozone pollutant effects on the respiratory function of children in southwest Mexico City. *Epidemiology* 1999; 10:8–16.

68. Kinney PL, Lippmann M. Respiratory effects of seasonal exposures to ozone and particles. *Arch Environ Health* 2000; 55:210–216.

69. Hackney JD, Thiede FC, Linn WS, Pedersen EE, Spier CE, Law DC, et al. Experimental studies on human health effects of air pollutants: IV. Short-term physiological and clinical effects of nitrogen dioxide exposure. *Arch Environ Health* 1978; 33:176–181.

70. Golden J, Nadel JA, Boushey HA. Bronchial hyperirritability in healthy subjects after exposure to ozone. *Am Rev Respir Dis* 1978; 118:287–294.

71. Holtzman MJ, Cunningham JH, Sheller JR, Irsigler GB, Nadel JA, Boushey HA. Effect of ozone on bronchial reactivity in atopic and nonatopic subjects. *Am Rev Respir Dis* 1980; 122:17–25.

72. Environmental Protection Agency. *National air quality and emission trends,* 1987. Report No. EPA/450/4–89/001. Washington, DC: Environmental Protection Agency, 1987.

73. Kuntasal G, Chang TY. Trends and relationships of ozone, NO_x, and HC in the south coast air basin of California. *J Air Pollut Control Assoc* 1987; 35:1158–1163.

74. National Research Council Committee on Medical and Biological Effects of Environmental Pollutants. *Carbon monoxide.* Washington, DC: National Academy of Sciences, 1977.

75. Coburn RF, Forster RE, Kane PB. Considerations of the physiological variables that determine blood carboxyhemoglobin concentration in man. *J Clin Invest* 1965; 44:1899–1910.

76. Winter PM, Miller JN. Carbon monoxide poisoning. *JAMA* 1976; 236:1502–1504.

77. Longo LD. The biological effects of carbon monoxide on the pregnant woman, fetus, and newborn infant. *Am J Obstet Gynecol* 1977; 129:69–103.

78. Zimmerman SS, Truxal B. Carbon monoxide poisoning. *Pediatrics* 1981; 68:215–224.

79. Astrup P, Kjeldsen J, Wanstrup J. Effects of carbon monoxide exposure on arterial walls. *Ann N Y Acad Sci* 1970; 174:294–300.

80. Theodore J, O'Donnell RD, Back KC.Toxicological evaluation of carbon monoxide in humans and other mammalian species. *J OccupMed* 1971; 13:242–255.

81. Sarma JS, Tillmanns H, Ikeda S, Bing RJ. The effect of carbon monoxide on lipid metabolism of human coronary arteries. *Atherosclerosis* 1975; 22:193–198.

82. Armitage AK, Davies RF, Turner DM. The effects of carbon monoxide on the development of atherosclerosis in the white carneau pigeon. *Atherosclerosis* 1976; 23:333–344.

83. El-Attar OA, Suiro DM. Effect of carbon monoxide on the whole fibrinolytic activity. *Ind Med Surg* 1968; 37:774–777.

84. Haft JI. Role of blood platelets in coronary artery disease. *Am J Cardiol* 1979; 43:1197–1206.

85. Schwartz J. Air pollution and hospital admissions for cardiovascular disease in Tucson. *Epidemiology* 1997; 8:371–377.

86. Schwartz J. Air pollution and hospital admissions for heart disease in eight U.S. counties. *Epidemiology* 1999; 10:17–22.

87. Samet JM, Marbury MC, Spengler JD. Health effects and sources of indoor air pollution. Part I. *Am Rev RespirDis* 1988; 137:221–242.

88. *Epidemiological, social and technical aspects of indoor air pollution from biomass fuel.* Geneva: World Health Organization, 1991.

89. Smith KR, Samet JM, Romieu I, Bruce N. Indoor air pollution in developing countries and acute lower respiratory infections in children. *Thorax* 2000; 55:518–532.

Water Quality and Water Resources

John Balbus

3

Safe drinking water is a fundamental human need. The sustainability of human society depends on the availability of plentiful and potable water. In the twenty-first century, our water supply faces a combination of old threats from human and industrial wastes and new threats from demand overload and climate change. Because of the different disciplines involved, previous reviews and texts have separately addressed the issues of water quality and water quantity.[1–6] This chapter integrates the challenges of maintaining water quantity as well as quality. Creative approaches to fulfilling human needs for water will require greater efforts to simultaneously address the questions of "how clean" and "how much."

Introduction

Globally, most of the world still endures the scourge of waterborne diseases that were familiar to the cities of the Western world during the nineteenth century. Diarrheal deaths, most due to waterborne infectious agents, kill over 2 million children annually.[7] Diarrheal illnesses, as many as a third of which are due to drinking contaminated water, sap the strength of 4 billion annually.[7] While the widespread use of oral rehydration therapy has substantially reduced the mortality associated with waterborne diseases, intensive efforts to provide improved water and sanitary infrastructure in developing nations have been less successful. Between

1990 and 2000, 800 million people gained access to water infrastructure, yet the proportion of the world's population without access remained near 20 percent, with an estimated 1.1 billion people without improved drinking water supplies.[8] Approximately 40 percent of the world's population, or 2.4 billion people, lack access to improved sanitary infrastructure.[8] Growing evidence indicates that updating the means of disposal of human wastes improves human health to a greater extent than purveying clean water supplies.[9]

Sources of Water

Drinking water is obtained primarily from surface water, such as streams, rivers, and lakes, or from groundwater through artesian or other types of wells. Other means of obtaining drinking water include collecting rainwater and desalinating seawater. The relative availability, treatment requirements, and most likely contaminants vary depending on the source of the water. Surface water is, in general, more prone to contamination by both microbial pathogens and chemicals, since it directly receives industrial and municipal wastewater and runoff from the land. Ground water is usually less contaminated than surface water because the soil through which it percolates serves as a filter. Exceptions occur when there is a direct or short path from the surface to the groundwater, when concentrated wastes deep in the soil leach into groundwater, as beneath a landfill, or when wastes are injected directly into the ground. Because groundwater generally collects slowly and does not quickly replenish itself, it remains contaminated for a long time once chemical wastes enter it.

Uses of Water

Water is used for a variety of purposes by developed societies; human consumption is actually one of the smallest uses. Worldwide, approximately 70 percent of the total withdrawal of water is used for irrigating agricultural land. The proportion of withdrawal used for irrigation varies widely by region, ranging from 88 percent in Africa to 31 percent in Europe.[10] (See table 3.1.) Industrial uses are substantial in the developing world; 55 percent of the water used in Europe and 45 percent in the United States go to industry. Most of this water is used for cooling or

Table 3.1 Worldwide freshwater withdrawals, 1970–1998

Continent	Year of data	Annual withdrawals: Total (cubic km)	Annual withdrawals: Per capita (cubic meters)	Sectoral withdrawals (percent): Domestic	Industrial	Agricultural
World	1987	3,240.00	645	8	23	69
Africa	1995	145.14	202	7	5	88
Europe	1995	455.29	625	14	55	31
North America	1991	512.43	1,798	13	47	39
South America	1995	106.21	335	18	23	59
Asia	1987	1,633.85	542	6	9	85
Oceania	1995	16.73	591	64	2	34

flushing equipment. It is thus generally returned to streams, rivers, or coastal waters, often either warmed or contaminated by chemicals. Within the home, far more tap water is used for flushing toilets, washing clothes, showering, and in many places watering lawns and washing cars than for drinking and cooking.[11] (See figure 3.1.)

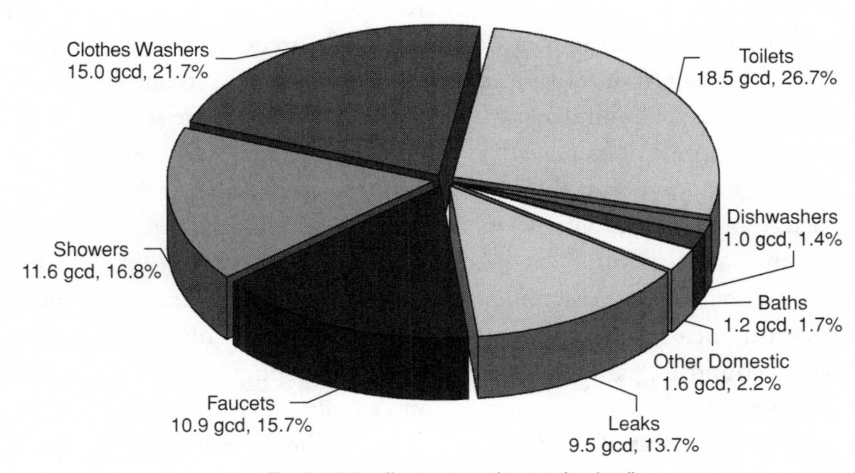

Figure 3.1 Mean per capita residential indoor water use

Sources of Contamination

Contaminants can enter drinking water by a variety of mechanisms. Many chemicals, such as pesticides and fertilizers, enter drinking-water sources after being deliberately applied to the land (or sometimes ponds and streams) and washing into surface or ground waters. Animal wastes from farms and wilderness areas enter water in a similar fashion. Industrial-waste chemicals and municipal wastes containing human fecal material and whatever chemicals have entered the system may be directly discharged into surface water, especially in developing countries. Other chemicals reach drinking-water sources after moving through soils. For example, cadmium and other heavy metals can leach out of dumps and landfills, while gasoline products and additives can contaminate drinking water through leakage from storage tanks. Some toxicants, such as arsenic and radon, occur naturally in the earth's crust and contaminate groundwater that is in contact with deposits of these substances. Lastly, our methods of treating and distributing drinking water lead to some chemicals (e.g., disinfection by-products and lead) being present within our drinking-water supplies.

Sources of contaminants that discharge into receiving waters from a pipe or other identifiable device are called *point sources*. Point sources include industrial wastes and sewage treatment plants. Sources of contaminants that cannot be defined by discrete pipes or other devices are called *non–point sources*. Examples of non–point sources include agricultural runoff, the runoff of oils and other chemicals from streets and highways, the multitude of leaking septic systems and gasoline storage tanks, and airborne emissions of mercury and other substances that may settle on and dissolve in source waters for drinking purposes. In the past 30 years, the passage of environmental legislation, such as the U.S. Clean Water Act and the Safe Drinking Water Act, has made a tremendous impact in reducing point-source contamination in the United States and other countries. But the existence of environmental regulations is often not enough. Many countries, particularly those in the developing world, continue to have serious problems with point-source contamination of water due to poor enforcement of existing laws.[12] Despite the success of regulations in reducing point-source contamination, however, drinking water remains vulnerable to contamination from non–point sources.

Water Treatment Process

The developed world reduced the burden of waterborne disease outbreaks by implementing a process to treat drinking water in the late nineteenth and early twentieth centuries. While the treatment of water has become more complex over the years, some of the process has changed little since its inception. First, water is pumped from its source. Surface waters containing sediment and microbial pathogens may be held in settling tanks or pretreated with chlorine at this stage. The next step involves adding a chemical coagulant, usually containing alum, to make small particles (such as pieces of decaying organic matter or microbes) stick together in the water. With flocculation, the collecting particles are stirred to make them adhere and become larger and heavier. This makes the particles more likely to sink to the bottom of holding tanks in the next step, sedimentation. These steps are meant to remove tiny particles, including microbes, from the water, but they do not remove chemicals to a significant degree. Water with chemical contamination may need to undergo additional treatment at this stage, such as activated charcoal or special ion exchange resins. Finally, prior to distribution to homes and businesses, the water is disinfected, usually with some form of chlorine. Alternative disinfectants include ozone and ultraviolet light. Chlorine or chloramine is still usually added when these alternatives are used; this is the only way to maintain a level of disinfecting agent in the water as it is distributed through the pipes and reservoirs to consumers. As the demand for higher-quality water increases, higher levels of technology, such as reverse osmosis and other membrane filtration techniques, are becoming more common.[13]

Treatment techniques in the developing world, in addition to the standard water treatment process described above, often rely on point-of-use methods performed in individual homes. Boiling water is the traditional method of water purification, but since it requires substantial energy production, its usefulness is limited. Boiling water also involves risks of scalding and lung injury when indoor cooking fuels are used. Iodine tablets have been used for many years. Newer techniques include the use of dark plastic bottles that collect solar energy to disinfect water[14,15] and the combination of chemical disinfectants and special narrow-necked storage bottles that prevent recontamination of water.[16,17] Both types of water treatment techniques are only effective against microbial pathogens and do not reduce the concentrations of chemical contaminants in the water.

Biological Threats to Water Quality

The World Health Organization (WHO) has described five ways water may be associated with human diseases. *Waterborne* diseases are transmitted by microorganisms that survive within water and are directly ingested. *Water-washed* diseases are those exacerbated by inadequate washing of hands or food, such as trachoma, other skin and eye infections, and many of the fecal-oral pathogens that cause waterborne disease. *Water-based* diseases are caused by organisms that spend part of their life cycle as larval forms within freshwater and come in contact with humans through bathing or ingesting of infested water. Examples include schistosomiasis and dracunculiasis. *Water-related* diseases, such as dengue fever and malaria, are caused by organisms that breed or otherwise spend part of their life cycle in water but do not come in contact with human hosts via water exposure. The last category, *water-dispersed* infections, are caused by microorganisms, such as *Legionella,* that proliferate within water supplies and are transmitted to humans by dispersal in air and inhalation.[18]

The majority of waterborne pathogens infect the gastrointestinal tract. They enter the water through contamination by human or animal fecal material. Microbial pathogens are usually categorized as bacterial, viral, or protozoal. These different categories of pathogens differ in their life cycles, their hosts and survival in the environment, and their removal by water treatment. A few microbial pathogens, such as *Legionella* and *Mycobacterium avium* species, have the capacity to grow within the pipes of water distribution systems. Most, however, contaminate the source water from which drinking water is obtained. To monitor the quality of source and drinking water, measurements of coliform bacteria (related to *E. coli)* and of water turbidity are commonly used. Although these methods provide a quick indication of fecal contamination of the water, they have been shown to be an unreliable safeguard. Tests for coliform bacteria will miss contamination by viruses and protozoa, which may be hardier than bacteria. In addition, outbreaks have occurred despite water turbidity within usual standards,[19,20] while warnings issued on the basis of increased turbidity have subsequently not been confirmed by evidence of microbial contamination.[21]

In both the developing and developed worlds, the burden of waterborne disease depends on the level of contamination of the source water and the availability and effectiveness of water treatment or alternative sources of water. When contamination is relatively continuous and treatment processes not effective, endemic disease rates are high. This is the case for bacterial pathogens in much of the developing world and probably for viral (e.g., Norwalk viruses) and protozoal (e.g., *Cryptosporidium parvum and Giardia lamblia*) pathogens in the developed world, although there are little data to evaluate this. The frequency and circumstances of outbreaks of waterborne infectious disease in the United States have changed in the past decade. Most outbreaks now are due to consumption of groundwater that is in contact with contaminated surface water but is inadequately treated.[22] In the past, waterborne outbreaks were slightly more frequent and more often due to consumption of surface water whose usual treatment process had failed.

Worldwide, the main bacterial pathogens spread by water are *Salmonella* species, *Shigella* species, pathogenic *E. coli,* and *Vibrio cholerae.* All of these bacteria cause diarrheal disease, although several cause disease with distinct features. *Salmonella typhi,* for example, causes typhoid fever, which is characterized by high fevers, abdominal pain, and more often constipation than diarrhea. *Shigella* and several other bacteria can cause dysentery, which is characterized by more severe abdominal cramping with blood and mucus in the stool. Different subspecies of *E. coli* can cause diarrhea with a variety of features and complications, including renal failure and coagulation disorders (associated with *E. coli* O157:H7). Lastly, the disease cholera is characterized by profuse, life-threatening passage of "rice-water" stools.

In general, bacterial pathogens are more effectively controlled than viral or protozoal pathogens by existing water treatment processes. Most bacteria are susceptible to inactivation by chlorine. The effectiveness of modern water treatment for bacterial pathogens is underscored by the low levels of waterborne bacterial diseases in developed countries and by the increasing proportion of outbreaks due to protozoal pathogens. For example, since 1980, there have been no outbreaks of bacterial waterborne disease related to disinfected water supplies.[22,23]

With the development of improved techniques for laboratory identification, viral pathogens have been shown to play an important role in waterborne gastrointestinal diseases. Outbreaks that have been characterized as "acute gastroenteritis of unknown etiology" are felt to be mostly due to viruses. The main waterborne viral pathogens are Norwalk and norwalk-like viruses, enteroviruses, and adenovirus types 40 and 41. In addition, viruses not commonly associated with water, such as hepatitis A and rotavirus, have had documented waterborne spread.[24–26] With the exception of hepatitis A, which causes liver inflammation, most of the viruses cause diarrheal illness. Most of the clinical illness caused by viruses is also acute and self-limited, with the possible exception of Coxsackie B virus, which has been associated with the chronic outcomes of myocarditis and diabetes mellitus.[27]

Due to their small size, viruses pass through both soil and water treatment filters readily. Thus, they are able to contaminate groundwater and to evade the water treatment process to some degree. Viruses vary in their susceptibility to chlorination. Rotaviruses, for example, are relatively susceptible to chlorination, while adenoviruses and hepatitis A virus are not as affected.[23]

Protozoa

Since the 1970s, a growing proportion of outbreaks of infectious origin in the United States have been due to protozoal pathogens, primarily *Giardia lamblia* and *Cryptosporidium parvum.* Both pathogens cause a spectrum of disease, from asymptomatic infection to chronic and severe diarrhea. *C. parvum,* in particular, causes severe and often fatal diarrheal disease in severely immunocompromised people, such as AIDS or cancer chemotherapy patients. Background endemic transmission of these pathogens is suggested by high rates of seropositivity (having antibodies to a specific pathogen) in the absence of known outbreaks. In the United States, studies have shown that 17 to 58 percent of adolescents and adults were seropositive for cryptosporidium, whereas in developing countries like Brazil and China, over half of the population is exposed by early childhood.[28,29]

Protozoal pathogens have a complex life cycle in which the infectious form for humans is a tiny, encapsulated cyst or oocyst. These forms are very hardy in the environment, able to survive for a long period in water or soil, and resistant to chlorination. Their ability to survive chlorination and to cause disease after the ingestion of a small number of oocysts makes protozoa a threat, even in treated water. In fact, large outbreaks of cryptosporidiosis have occurred in the United States in communities (Milwaukee, Wisconsin; Las Vegas, Nevada; and Carrollton, Georgia) served by modern water utilities and with water that continuously met federal standards.[19,20,30] These outbreaks have prompted U.S. government regulators to require more stringent water treatments that address the threat of protozoal pathogens.[31,32]

Chemical Threats to Water Quality

In addition to biological agents, several types of toxic chemicals are common drinking water contaminants.

Nitrates

Nitrates can accumulate in ground and surface water, both from fertilizer runoff and also from leaching of ammonia compounds from septic tanks and other waste systems.[33] After ingestion, nitrates are converted into more toxic nitrites by bacteria colonizing the gastrointestinal tract. Nitrites oxidize the hemoglobin in red blood cells and can lead to methemoglobinemia, which inhibits the red blood cell's ability to transfer oxygen to the rest of the body. Methemoglobinemia from ingestion of nitrates in drinking water occurs almost exclusively in children under four months because they do not have enough of the hemoglobin reductase enzyme that restores hemoglobin to its functional form.[34] Methemoglobinemia results in oxygen starvation, cyanosis (bluish discoloration of the skin and lips), and, if severe enough, brain damage and death.

Deaths from nitrate poisoning are rare but still occur in the United States and elsewhere, predominantly in rural agricultural areas where private wells are used for drinking water. Wells that are particularly shallow and have leaking casings are at highest risk of accumulating large amounts of nitrate. Most states and counties in the United States have programs to test private wells for nitrate levels to identify families at risk. A United Nations study suggested that nitrate contamination of drinking water will be an increasingly severe problem for North America and Europe. While developing countries' level of nitrate contamination is generally lower, the study indicated that India and Brazil have growing problems with nitrate levels in their water.[35]

Nitrate exposure may also cause adverse reproductive outcomes, including spontaneous abortion and neural tube defects.[36–38] In one report, each of the four women experiencing spontaneous abortions had normal pregnancies preceding or following the period of consuming the nitrate-contaminated drinking water. Levels of nitrates in the wells of the affected women ranged from 19 to 26 mg/L, well above the EPA standard of 10 mg/L.[36] Despite this one case study, the association between nitrate exposure and adverse reproductive outcomes has not yet been adequately studied to conclude a causal relationship.

Pesticides

Because of their application in enormous amounts to both agricultural and managed urban landscapes, pesticides are commonly found in both groundwater and surface water. In the United States and other developed countries, the majority of chemicals applied to the land are herbicides. This is part of the reason that herbicides are more commonly found than insecticides in U.S. water supplies.[39] In the developing world, limited resources lead to less overall use of chemical pesticides, with a greater proportion of insecticides used than herbicides.

Many of the most harmful pesticides to humans, wildlife, or both have been banned for use in the United States over the past 30 years. Because many of these are persistent in the environment, however, they can still be found in small amounts in drinking water (and in humans). Thus, insecticides like DDT and dieldrin are still present in source waters.[39] More important, many chemical pesticides now banned in the United States are still being used throughout the rest of the world, particularly in poorer countries, and are likely to constitute a global health threat through drinking-water contamination.

There are no reports of acute pesticide poisoning due to drinking water contamination, but severe contamination events have led to widespread closure of wells. For example, intensive use of the nematocide dibromochloropropane (DBCP) in the Central Valley of California during the 1950s, 1960s, and 1970s resulted in contamination of at least 2,500 wells. Agricultural workers exposed to DBCP suffered permanent sterility, but no adverse reproductive effects were found in the population exposed through their drinking water.[40] Similar contamination of drinking-water sources with DBCP and aldicarb has occurred in Hawaii, Arizona, South Carolina, Maryland, and Long Island, New York.[41]

In addition to concerns about health effects of severe contamination is a concern about the chronic health effects of long-term exposure to pesticides. Commonly used herbicides, such as simazine and atrazine, have little acute toxicity for humans, but may have subtle effects on various endocrine systems, potentially causing adverse outcomes if exposure occurs during critical stages of human development. A number of pesticides, including the herbicide 2,4-D and organochlorine insecticides such as DDT and dieldrin, have been associated with several different types of cancer. Most of the studies showing an association were of occupationally exposed populations. It is unclear whether populations exposed to pesticides at lower levels through water consumption have any increased cancer risk from such exposure.[42]

Industrial Discharges

Contamination of water by industrial chemicals probably peaked in the United States during the 1960s and 1970s. Events such as the 1969 fire on the Cuyahoga River in Ohio, in which a spark ignited accumulated debris and an oil slick on the river, focused the public's attention on the deteriorating quality of the nation's water. An influential study conducted by the U.S. EPA in 1972 found 36 organic compounds in the waters of the lower Mississippi. In response, Congress enacted new, strict federal regulations, including the Clean Water Act of 1972 and the Safe Drinking Water Act of 1974. These regulations have substantially reduced the discharge of industrial chemicals directly into drinking-water sources. Contamination still occurs, however, through the seepage of chemicals from underground storage tanks, old hazardous waste sites, and contaminated sediments.

The widespread contamination of water supplies with methyl-tertiary-butyl-ether (MTBE) is an example of the impacts of leaking underground storage tanks. MTBE is a gasoline additive designed to lower toxic air emissions from automobile exhausts and has constituted up to 15 percent of gasoline in certain regions.[43] It surprised scientists by moving very quickly and persistently through the soil into groundwater, leading to widespread and occasionally highly concentrated contamination of wells.

It also can enter water supplies to a lesser extent through air releases. The toxicity of MTBE in drinking water has not been thoroughly investigated; most toxicology studies have used inhalation as the route of exposure.[44] It has been estimated, however, that the health risk from low levels of MTBE in drinking water is insignificant.[43]

The Agency for Toxic Substances and Disease Registry (ATSDR) has characterized drinking-water contamination associated with Superfund sites in the United States. Among the most commonly found chemicals were trichloroethylene and perchloroethylene, chlorinated solvents used widely in industry.[45] These chemicals in high doses are associated with liver and kidney damage, neurotoxicity, and possibly cancer in the case of tetrachloroethylene. Due in part to the difficulty of measuring exposures to low levels and mixtures of industrial chemicals, there are very few data to assess the health effects of these chemicals in drinking water.

Radon

Radon dissolves in water as it moves through cracks in the earth's crust. Because it dissipates when the water comes in contact with air, it is only found in significant concentrations in groundwater that is encased in solid clay or rock; surface waters do not pose any risk. Radon is not human-made but occurs where uranium deposits are found within the earth's crust. It emits alpha particles that can cause deletions, rearrangements, and other damage to DNA at the site of exposure. Although concentrations of 10,000,000 Bq/m^3 (about 270,000 pCi/L) or more of radon are known to exist in drinking water, the average concentration in drinking water derived from groundwater sources is about 20,000 Bq/m^3 (540 pCi/L).[46] Radon volatilizing from drinking water is estimated to raise indoor air concentrations by one ten-thousandth of the concentration of radon in the water. Thus, a radon concentration of 270,000 pCi/L in water would be expected to raise the concentration in indoor air by 27 pCi/L. Homes with an average level of radon in the drinking water would have a negligible increase in indoor air radon concentration. Areas within the United States with naturally high radon deposits include the Appalachian regions, New England, and some of the Rocky Mountain and Southwest regions.[46]

High levels of radon in household water may slightly increase the risk of lung cancer from inhaling aerosolized water and, to a lesser degree, increase the risk of gastric cancer through direct ingestion. A National Academy of Sciences (NAS) report estimated the lifetime risk of cancer (mainly gastric cancer) owing to ingesting water with radon concentrations of 1 Bq/m^3 to be 0.2×10^{-8}. The lifetime risk of lung cancer in a mixed population of smokers and nonsmokers (from aerosolized radon) was estimated at 1.6×10^{-8}. These risk estimates corresponded to approximately 180 annual deaths from radon in drinking water per year in the United States. After comparing this to estimates of deaths from radon in indoor air that were 100 times greater, the NAS committee concluded that public health efforts should focus on mitigating air concentrations unless water concentrations of radon were extremely high. Nonetheless, these calculated risks for radon were large compared to the risks of other regulated chemical contaminants in drinking water.[46]

Arsenic

Arsenic generally appears in local water supplies due to the leaching of deposits of arsenic in the earth's crust, but it is also a by-product of several industries, including pesticides, wood preservatives, and mining. High levels of arsenic in drinking water have been associated with several adverse health effects, including cancer of the skin, lung, and bladder, skin hyperpigmentation and keratosis, vascular disease, and neurotoxicity. Parts of the world with the highest arsenic levels include the Bengal area of India and Bangladesh, Taiwan, Argentina, and Chile. Within the United States, arsenic levels are highest in the Western states, Upper Midwest, and New England. The current WHO standard for arsenic in drinking water is 10 μg/L. At this writing, the United States is debating lowering its standard for arsenic from the level of 50 μg/L, which was set in 1942, to somewhere between 3 and 20 μg/L.

The problem in Bengal and Bangladesh is illustrative of the complex and difficult choices faced by developing countries. The governments of India and Bangladesh promoted the digging of tube wells from 1950 through the 1970s to provide an alternative drinking water source to the severely contaminated surface water in the region. The tube wells also provided abundant and consistent water for irrigation, and soon the area was producing four rice crops instead of the single crop when rain was relied on for irrigation. In the 1980s, it became apparent that the water was contaminated by arsenic, and hundreds of thousands of people developed skin manifestations of arsenic poisoning. Abandoning the wells would mean reverting to surface waters with heavy biological contamination and significant economic losses due to lost crop yields. Using the wells safely will require expensive water treatment devices.

Lead

The use of lead for water distribution pipes dates back to Roman times; thus, the latin word for lead, *plumbum,* is the source of the English word *plumbing*. Depending on its mineral characteristics, water passing through lead pipes will dissolve the lead to a greater or lesser extent. Water that is hot and soft dissolves more lead than cold, hard water. Lead was used in water mains in cities during the late 1800s, but very few of these are still in service. Lead pipes have not been used in homes since the early 1900s, and the lead content of the solder used to join copper pipes has been significantly reduced since 1986. Nonetheless, lead is still present as a component of brass fixtures and, in many cases, in the soldering used in water fountains and other faucets or fixtures. Because water that has sat in pipes and fixtures within the home for a prolonged period is more likely to accumulate lead, it is a good idea to let water run before consuming it if the tap has not been used for a long time. The EPA estimates that approximately 20 percent of the background exposure to lead in the general population comes from drinking water.[47]

Lead exposure, even in relatively small concentrations, is associated with decreased IQ scores and behavioral problems in infants and children[48–50] and with hypertension in adults.[51] The current standard for lead in drinking water is 15 μg/L.[52] Because the source of lead in drinking water is almost always within the home rather than at the source or in the treatment plant, monitoring a water system for lead requires sampling the taps of individual homes. The current standard requires that no more than 10 percent of homes sampled may exceed the standard concentration.

Disinfectant By-Products

Disinfectant by-products are formed by the reaction of dissolved hydrocarbons with chlorine, which is generally added for disinfection, and occasionally bromine, which occurs naturally in some water sources. The hydrocarbon compounds come primarily from decayed vegetation and dead microorganisms. The most common DBPs are the trihalomethanes (THMs) and haloacetic acids (HAAs).[23] Growing epidemiologic evidence suggests that consumption of this complex mixture of chemicals in drinking water is associated with cancers of the bladder[53,54] and colon[55] and with adverse reproductive outcomes[56–58]

The strongest evidence exists for an association between trihalomethanes and cancer of the bladder. The EPA has estimated that 1,100–9,300 cases of cancer could be prevented annually by lowering the standard for trihalomethanes from 100 to 80 parts per million.[59]

Threats to Water Quantity or Availability

In the future, water quantity and availability are likely to become as critical to human well-being as water quality and purity. The total amount of freshwater available for human use is limited: roughly 12,500 cubic kilometers, or 0.007 percent of all the water on earth.[35] At present, humans use roughly half of the amount of water available. Growing populations and economic development will lead to greater consumption, putting a severe strain on aquatic ecosystems and jeopardizing other uses of freshwater such as recreation and navigation. Water availability may also be threatened by climate changes that alter the seasonal patterns of precipitation on which communities depend for their water supply or closure of wells or surface water pumps due to chemical contamination of the water source.[60]

The balance between growing population demand and water supplies has been narrowing severely in many places, both in the United States and around the world. Roughly 8 percent of the world's population lives in areas considered to be highly water stressed, meaning more than 40 percent of available water resources are currently being consumed; another quarter of the world's population lives in areas that are moderately water stressed, or consuming between 20–40 percent of available water.[35] (See figure 3.2.)

During the next century, changes in global climate are likely to have a profound affect on water supplies in certain regions. Overall stream flows are likely to decrease in several areas that are currently water-stressed, such as Central Asia, Mediterranean countries, and southern Africa. On the other hand, several areas are likely to see an increase in water availability, including Southeast Asia and higher-latitude countries.[60]

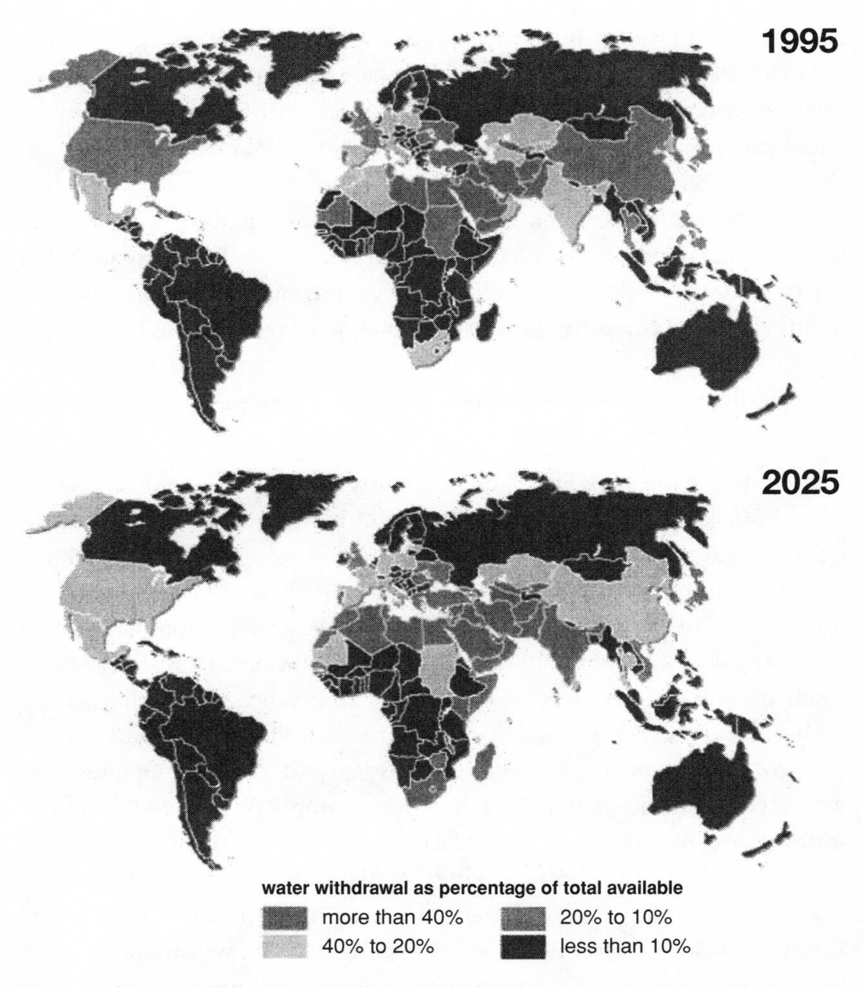

Figure 3.2 (*Top*) Water withdrawal as a percentage of water availability—1995; (*bottom*) water withdrawal as a percentage of water availability—2025

In addition, the timing and amount of runoff from mountain snowpacks are very sensitive to changes in precipitation and temperature. While predictive models are not consistent as to whether precipitation will increase or decrease in many mountainous areas, they are consistent in showing that increased temperatures will lead to increased winter runoff and decreased spring and summer runoff, thus reducing the availability of water for much of the year. Areas that depend on mountain snowpack for their drinking water, such as the Bay Area of California, will have to adapt.[61]

Threats to water availability may lead to increased tension within and between nations in various areas. The past 50 years have seen conflicts over water resources arise between India and Pakistan, Egypt and Sudan, and Turkey and Iraq, and among Israel, Jordan, and Syria.[62] Nearly half the world's land surface serves as the watershed for one of 261 rivers that receive the runoff of more than one nation.[62] In recent history, armed conflict over water resources has been rare, and water shortages have often led to innovative and cooperative solutions. Nonetheless, a greater frequency and severity of water shortages may cause further deterioration of international relations already strained for other reasons, in areas such as the Middle East or Southern Asia.[63]

Solutions for Water Quality

The difficulty of providing clean and safe drinking water is related to the degree of contamination of the source from which the water is drawn. Solutions can be divided into either technological innovations capable of cleaning more contaminated water to a greater level of purity, or prevention of source-water contamination in the first place.

Standard water treatment techniques do not reduce the concentration of chemicals in the water to a significant degree, and they also do not always remove or inactivate microbial pathogens to a sufficient degree to prevent disease from occurring. More complicated (and more expensive) treatment techniques are capable of addressing these problems. The selection of the treatment process and the use of advanced technologies will depend on the particular problem being addressed. For example, technologies for chemical contamination include granulated activated charcoal (GAC), ion exchange resins, and reverse osmosis. GAC is useful for removing organic chemicals and certain inorganic chemicals, including chlorine. It is not effective at removing lead, nitrates, or other charged molecules. Ion exchange resins and other adsorptive treatments, such as activated alumina, on the other hand, are designed to remove charged molecules, such as lead, nitrates, and fluoride, as well as ions that contribute to water hardness, such as calcium and sodium. Reverse osmosis and nanofiltration processes use membranes that selectively block all molecules larger than about 0.05 microns and are able to exclude synthetic organic molecules to some extent.[64] Thus, the design of a treatment plan to remove chemical contaminants from drinking water needs to be tailored to the specific contaminants to be removed.

Due to the expense of these treatment techniques and the recognition of other ecological benefits, there has been increasing emphasis on watershed protection, both in water regulations and in the practices of a growing number of communities. New York City, for example, has obtained its water for years from a series of lakes and reservoirs located in the rural Catskill area of Upstate New York and uses this water unfiltered. As increasing development has occurred in the watersheds of these reservoirs and more stringent water-quality standards have been put in place, New York City has needed to manage more actively the watershed that provides drinking water to its 9 million residents. The only alternative to watershed protection is a new, costly filtration system. A memorandum of agreement put into place in 1997 committed the city to a massive program of land purchase, water-quality monitoring, disease surveillance, and economic aid to the communities whose development might be restricted under the watershed management plan.[65] The New York City story is a dramatic illustration of prevention versus control policy options.

New drinking-water regulations in the United States, such as the Safe Drinking Water Act (SDWA) Amendments of 1986 and 1996, have placed greater emphasis on watershed protection as well. The SDWA Amendments require and provide funding for each state to produce a State Watershed Assessment Plan (SWAP), which identifies critical watershed areas and analyzes the use of land within that watershed.

Given the probability of human or mechanical failure in technological solutions to contaminated water and the other ecological benefits of watershed protection, the quality of drinking water should not depend solely on technological solutions. The sustainability and availability of clean drinking water in the future will depend on careful management of drinking-water sources, not only within the water treatment plant, but also well before the water reaches the plant.

Solutions for Water Quantity

The provision of adequate clean water in the future will depend on acts that increase water quantity as well as quality. As new, untapped sources of water become more difficult to find and existing groundwater sources are withdrawn at rates near or beyond their capacity to recharge, innovations will be required to reduce the amount of water used per person without jeopardizing the health of the people or the economic viability of communities. Many communities around the world have had to find ways to provide adequate water supplies in the face of extreme scarcity and have developed a variety of approaches. Improved efficiency of existing uses, reclamation and reuse of wastewater for specific uses, and altered demand are some of the ways water quantity can be preserved or increased.

Because water has generally been considered an endless resource, most of the world's uses of water have not been designed for efficiency. As a result, tremendous amounts of water are wasted through evaporation, leakage, or mismatch between the amount required for the purpose and the amount used. Since around two-thirds of the world's water use is for irrigating crops, more efficient irrigation techniques can make a considerable impact on the world's use of water.

Much of the world uses canals and flooding techniques to irrigate agricultural land. With these techniques, much of the water withdrawn is either lost to evaporation or misdirected and does not reach the crops. The use of low-pressure drip-irrigation devices can improve efficiency of water use 30–70 percent, while also increasing crop yield.[66] Low-pressure sprinklers can also lead to highly efficient use of water. Unfortunately, because of higher costs of these methods and policies that do not promote more efficient use of irrigation water, only 10–15 percent of the world's cropland is irrigated by sprinklers and less than 1 percent by drip devices. Further improvement in efficiency can be obtained by using computer models to help decide exactly how much and when to apply irrigation water.[66] Clearly, part of the solution needs to be promoting policies and investments that facilitate the adoption of these more efficient irrigation methods.

There is also much room for improvements in the efficiency of domestic water use. Within the home, low-flow toilets and showers can reduce water consumption by significantly. New York City, in the face of a water shortage in the early 1990s, promoted replacement of standard toilets that used 6 gallons per flush with low-flow toilets using only 1.6 gallons per flush. As a result, water use per building per year decreased 29 percent.[67] Within cities, there is also a great need to monitor and replace leaking water distribution pipes, which can lose as much as 20 percent of the water leaving the treatment plants into the ground before reaching consumers' taps.

Much of the domestic use of the home is for toilet flushing, car washing, lawn watering, and other uses that do not require water fit for drinking. Another effective way to make more of the water resources available is to reclaim and reuse wastewater and provide it for these types of uses that do not require the highest water quality. For years, the capital city of Windhoek, Namibia, located in a water-scarce region, has been using reclaimed water in its parks and for irrigation. Even though the reclaimed water is measured to be of as high or higher quality than the city's drinking water, public acceptance of the reclaimed water is low. Nonetheless, the city has had to rely on this water during times of extreme drought, as in 1995, when it provided as much as 30 percent of the water supplied to private homes.

Lastly, reduced use of water per person can be achieved by changing human behaviors that lead to water consumption. For example, in the arid states of the Western United States, many communities have encouraged a change from irrigated-grass-lawn landscaping to more native plants that do not require as much water. A more fundamental and perhaps more difficult change in behavior would be changing food-consumption patterns throughout the world, particularly in the developed world. A preference for grain and vegetable consumption over animal meats would considerably reduce water demand, as well as make more land available for growing food for humans. It is estimated that 100 to 250 gallons of water are required to produce a pound of corn, while 2000 to 8500 gallons are needed to grow the grain to produce a pound of beef.[62]

The way water is used and even thought about is evolving. Water-rich parts of the world can no longer take this resource for granted, and water-poor areas are under increasing pressure to find new ways to provide adequate water for all of their needs. Tremendous progress has been made in providing safe and healthful water to billions of people in both developed and developing countries. Innovation, money, and determination will be required to maintain the quality of water supplies and find ways to bring that water to all people.

References

1. Conway JB. Water quality management. In: Wallace RB, Doebbeling BN, Last JM, editors. *Public health and preventive medicine.* 14th ed. Stamford, CT: Appleton and Lange, 1998. pp. 737–763.

2. Gleick, PH. *Water: The potential consequences of climate variability and change for the water resources of the United States.* Oakland, CA: Pacific Institute for Studies in Development, Environment, and Security, 2000. ISBN #1–893790–04–5. pp. 1–151.

3. National Research Council. *Drinking water and health.* vol. 1. Washington, DC: National Academy of Sciences, 1977.

4. National Research Council. *Drinking water and health.* vol. 2. Washington, DC: National Academy Press, 1980.

5. National Research Council. *Drinking water and health.* vol. 3. Washington, DC: National Academy Press, 1980.

6. National Research Council. *Drinking water and health.* vol. 4. Washington, DC: National Academy Press, 1982.

7. Ford TE. Microbiological safety of drinking water: United States and global perspectives. *Environ Health Perspect* 1999; 107:191–206.

8. WHO/UNICEF Joint Monitoring Programme for Water Supply and Sanitation. 2000. *The global water supply and sanitation assessment 2000.* World Health Organization and United Nations Children's Fund. Available: http://www.who.int/water_sanitation_health/Globassessment/GlobalTOC.htm

9. Esrey SA. Water, waste, and well-being: A multicountry study. *Am J Epidemiol* 1996; 143:608–623.

10. World Resources Institute, United Nations Environment Programme, United Nations Development Programme, et al. *World resources, 1998–99: A guide to the global environment.* New York: Oxford University Press, 1998.

11. Mayer PW, DeOreo WB, Opitz EM, Kiefer JC, Davis WY, Dziegielewski B, et al. *Residential end uses of water.* Denver: American Water Works Association Research Foundation, 1999.

12. *Global environment outlook 2000.* New York: United Nations, 2000.

13. Logsdon GS, Hess A, Horsley M. Guide to selection of water treatment processes. In: Letterman RD, editor. *Water quality and treatment: A handbook of community water supplies.* 5th ed. New York: McGraw-Hill, 1999. pp. 3.1–3.26.

14. Conroy RM, Meegan ME, Joyce T, McGuigan K, Barnes J. Solar disinfection of water reduces diarrhoeal disease: An update. *Arch Dis Child* 1999; 81:337–338.

15. McGuigan KG, Joyce TM, Conroy RM. Solar disinfection: Use of sunlight to decontaminate drinking water in developing countries. *J Med Microbiol* 1999; 48:785–787.

16. Mintz ED, Reiff FM, Tauxe RV. Safe water treatment and storage in the home: A practical new strategy to prevent waterborne disease. *JAMA* 1995; 273:948–953.

17. Quick RE, Venczel LV, Mintz ED, Soleto L, Aparicio J, Gironaz M, et al. Diarrhoea prevention in Bolivia through point-of-use water treatment and safe storage: A promising new strategy. *Epidemiol Infect* 1999; 122:83–90.

18. World Health Organization. *Our planet, our health: Report of the WHO Commission on Health and the Environment.* Geneva: World Health Organization, 1992.

19. Goldstein ST, Juranek DD, Ravenholt O, Hightower AW, Martin DG, Mesnik JL, et al. Cryptosporidiosis: An outbreak associated with drinking water despite state-of-the-art water treatment. *Ann Intern Med* 1996; 124:459–468.

20. Hayes EB, Matte TD, O'Brien TR, McKinley TW, Logsdon GS, Rose JB, et al. Large community outbreak of cryptosporidiosis due to contamination of a filtered public water supply. *N Engl J Med* 1989; 320:1372–1376.

21. Centers for Disease Control. Assessment of inadequately filtered public drinking water: Washington, DC, 1993. *MMWR* 1994; 43:661–669.

22. Centers for Disease Control. Surveillance for waterborne-disease outbreaks—United States, 1997–1998. *MMWR* 2000; 49:1–21.

23. Cohn PD, Cox M, Berger PS. Health and aesthetic aspects of water quality. In: Letterman RD, editor. *Water quality and treatment: A handbook of community water supplies*. 5th ed. New York: McGraw-Hill, 1999. pp. 2.1–2.86.

24. Gerba CP, Rose JB, Haas CN, Crabtree KD. Waterborne rotavirus: A risk assessment. *Wat Res* 1996; 30:2929–2940.

25. Grabow WOK. Hepatitis viruses in water: Update on risk and control. *Water SA* 1997; 23:379–385.

26. Hopkins RS, Gaspard GB, Williams FP Jr, Karlin RJ, Cukor G, Blacklow NR. A community waterborne gastroenteritis outbreak: Evidence for rotavirus as the agent. *Am J Public Health* 1984; 74:263–265.

27. Gerba CP, Rose JB. Viruses in source and drinking water. In: McFeters GA, editor. *Drinking water microbiology: Progress and recent developments*. New York: Springer-Verlag, 1990. pp. 380–396.

28. Kuhls TL, Mosier DA, Crawford DL, Griffiths J. Seroprevalence of cryptosporidial antibodies during infancy, childhood, and adolescence. *Clinical Infectious Diseases* 1994; 18:731–735.

29. Zu SX, Li JF, Barrett LJ, Fayer R, Shu SY, McAuliffe JF, et al. Seroepidemiologic study of cryptosporidium infection in children from rural communities of Anhui, China and Fortaleza, Brazil. *Am J Trop Med Hyg* 1994; 51:1–10.

30. MacKenzie WR, Hoxie NJ, Proctor ME, Gradus MS, Blair KA, Peterson DE, et al. A massive outbreak in Milwaukee of cryptosporidium infection transmitted through the public water supply. *N Engl J Med* 1994; 331:161–167.

31. US Environmental Protection Agency. *National primary drinking water regulations: Interim enhanced surface water treatment*. 40 CFR (parts 9, 141, 142). Washington, DC: US Environmental Protection Agency, 1998 Dec 16.

32. US Environmental Protection Agency. *National primary drinking water regulations: Long term 1 enhanced surface water treatment and filter backwash rule—proposed*. 40 CFR (parts 141, 142). Washington, DC: US Environmental Protection Agency, 2000 Apr 10.

33. Centers for Disease Control. Methemoglobinemia in an infant—Wisconsin, 1992. *MMWR* 1993; 42:217–219.

34. Johnson CJ, Kross, BC. Continuing importance of nitrate contamination of groundwater and wells in rural areas. *Am J Ind Med* 1990; 18:449–456.

35. Commission on Sustainable Development. *Comprehensive assessment of the freshwater resources of the world.* Report of the Secretary-General. New York: United Nations Department for Policy Coordination and Sustainable Development, 1997 Feb 4. E/CN.17/1997/9.

36. Centers for Disease Control. Spontaneous abortions possibly related to ingestion of nitrate-contaminated well water—LaGrange County, Indiana, 1991–1994. *MMWR* 1996; 45:569–572.

37. Centers for Disease Control. Methemoglobinemia attributable to nitrite contamination of potable water through boiler fluid additives—New Jersey, 1992 and 1996. *MMWR* 1997; 46:202–204.

38. Croen LA, Todoroff K, Shaw GM. Maternal exposure to nitrate from drinking water and diet and risk for neural tube defects. *Am J Epidemiol* 2001; 153:325–331.

39. US Geological Survey. *The quality of our nation's waters: Nutrients and pesticides—A summary. USGS Circular 1225.* Reston, VA: US Geological Survey 1999; 116–199.

40. Whorton MD, Wong O, Morgan RW, Gordon N. An epidemiologic investigation of birth outcomes in relation to dibromochloropropane contamination in drinking water in Fresno County, California, USA. *Int Arch Occup Environ Health* 1989; 61:403–407.

41. Brown JP, Jackson RJ. Water pollution. In: Brooks S, Gochfeld M, Herzstein J, Schenker M, Jackson R, editors. *Environmental medicine.* St. Louis: Mosby, 1995. pp. 479–487.

42. Calderon RL. The epidemiology of chemical contaminants of drinking water. *Food Chem Toxicol* 2000; 38: Suppl 1:S13–S20.

43. Stern BR, Tardiff RG. Risk characterization of methyl tertiary butyl ether (MTBE) in tap water. *Risk Anal* 1997; 17:727–743.

44. US Environmental Protection Agency. *Oxygenates in water: Critical information and research needs.* EPA/600/R–98/048. Washington, DC: US Environmental Protection Agency, 1998. pp. 1–65.

45. Johnson BL, DeRosa CT. The toxicologic hazard of superfund hazardous waste sites. *Rev Environ Health* 1997; 12:235–251.

46. Committee on Risk Assessment of Exposure to Radon in Drinking Water, National Research Council. *Risk assessment of radon in drinking water, executive summary.* Washington, DC: National Academy Press, 1999.

47. US Environmental Protection Agency. *Drinking water priority rulemaking: Lead and copper rule-minor revisions.* 2000. Available: http://www.epa.gov/safewater/stadard/lead&co1.html

48. McMichael AJ, Baghurst PA, Vimpani GV, Wigg NR, Robertson NF, Tong S. Tooth lead levels and IQ in school-age children: The Port Pirie Cohort Study. *Am J Epidemiol* 1994; 140:489–499.

49. Needleman, HL, Riess JA, Tobin MJ, Biesecker GE, Greenhouse JB. Bone lead levels and delinquent behavior. *JAMA* 1996; 275: 363–369.

50. Needleman HL, Schell A, Bellinger D, Leviton A, Allred EN. The long-term effects of exposure to low doses of lead in childhood. An 11-year follow-up report. *N Engl J Med* 1990; 322:83–88.

51. Cheng Y, Schwartz J, Sparrow D, Aro A, Weiss ST, Hu H. Bone lead and blood lead levels in relation to baseline blood pressure and the prospective development of hypertension: The Normative Aging Study. *Am J Epidemiol* 2001; 153:164–171.

52. US Environmental Protection Agency. *Current drinking water standards*. 2001 Mar 2. Available: http://www.epa.gov/safewater/mcl.html

53. Cantor KP, Lynch CF, Hildesheim ME, Dosemeci M, Lubin J, Alavanja M, et al. Drinking water source and chlorination byproducts. I. Risk of bladder cancer. *Epidemiol* 1998; 9:21–28.

54. Freedman DM, Cantor KP, Lee NL, Chen LS, Lei HH, Ruhl CE, et al. Bladder cancer and drinking water: A population-based case-control study in Washington County, Maryland (United States). *Cancer Causes Control* 1997; 8:738–744.

55. Hildesheim ME, Cantor KP, Lynch CF, Dosemeci M, Lubin J, Alavanja M, et al. Drinking water source and chlorination byproducts. II. Risk of colon and rectal cancers. *Epidemiol* 1998; 9:29–35.

56. Swan SH, Waller K, Hopkins B, Windham G, Fenster L, Schaefer C, et al. A prospective study of spontaneous abortion: Relation to amount and source of drinking water consumed in early pregnancy. *Epidemiol* 1998; 9:126–133.

57. Waller KS, Swan H, DeLorenze G, Hopkins B. Trihalomethanes in drinking water and spontaneous abortion. *Epidemiol* 1998; 9:134–140.

58. Klotz JB, Pyrch LA. Neural tube defects and drinking water disinfection byproducts. *Epidemiol* 1999; 10:383–390.

59. US Environmental Protection Agency. National primary drinking water regulations: Disinfectants and disinfection byproducts notice of data availability. *Federal Register* 1998; 63:15674–15692.

60. Working Group II of the Intergovernmental Panel on Climate Change. *Climate change 2001: Impacts, adaptation, and vulnerability*. Cambridge, UK: Cambridge University Press, 2001.

61. Frederick KD, Gleick PH. *Water and global climate change: Potential impacts on US water resources*. Arlington, VA: Pew Center on Global Climate Change, 1999.

62. Gleick PH. Making every drop count. *Sci Am* 2001; 284(2):40–45.

63. Kennedy D, Holloway D, Weinthal E, Falcon W, Ehrlich P, Naylor R, et al. *Environmental quality and regional conflict*. Washington, DC: Carnegie Corporation of New York, 1998.

64. Taylor JS, Wiesner M. Membranes. In: Letterman, RD, editor. *Water quality and treatment: A handbook of community water supplies*. 5th ed. New York: McGraw-Hill, 1999. pp. 11.1–11.71.

65. Committee to Review the New York City Watershed Management Strategy, Water Science and Technology Board Commission on Geosciences Environment and Resources, National Research Council. *Watershed management for potable water supply: Assessing the New York City strategy*. Washington, DC: National Academy Press, 1999.

66. Postel S. Growing more food with less water. *Sci Am* 2001; 284(2):46–51.

67. Martindale D, Gleick PH. Safeguarding our water: How we can do it. *Sci Am* 2001; 284(2):52–55.

Human Health and Heavy Metals Exposure

Howard Hu

4

Metals, a major category of globally distributed pollutants, are natural elements that have been extracted from the earth and harnessed for human industry and products for millennia. (An exception to metals being "natural" is plutonium, the material at the heart of nuclear weapons, created by humans through the use of uranium as nuclear reactor fuel.) Metals are notable for their wide environmental dispersion from such activity; their tendency to accumulate in select tissues of the body; and their overall potential to be toxic even at relatively minor levels of exposure. Some metals, such as copper and iron, are essential to life and play irreplaceable roles in, for example, the functioning of critical enzyme systems. Other metals are *xenobiotics*—that is, they have no useful role in human physiology (and most other living organisms)—and, even worse, as in the case of lead and mercury, may be toxic even at trace levels of exposure. Even metals that are essential have the potential to become harmful at high levels of exposure, a reflection of a basic tenet of toxicology—"the dose makes the poison." One reflection of the importance of metals relative to other potential hazards is their ranking by the U.S. Agency for Toxic Substances and Disease Registry (ATSDR), which lists all hazards present in toxic waste sites according to their prevalence and the severity of their toxicity. The first, second, third, and sixth hazards respectively on the list are heavy metals: lead, mercury, arsenic, and cadmium.

Exposure to metals can occur through a variety of routes. Metals may be inhaled as dust or fume (tiny particulate matter, such as the lead oxide particles produced by the combustion of leaded gasoline). Some metals can be vaporized (e.g., mercury vapor in the manufacture of fluorescent lamps) and inhaled. Metals may also be ingested involuntarily through food and drink. The amount actually absorbed from the digestive tract can vary widely, depending on the chemical form of the metal and the age and nutritional status of the individual. Once a metal is absorbed, it distributes in tissues and organs. Excretion typically occurs primarily through the kidneys and digestive tract, but metals tend to persist in some storage sites, like the liver, bones, and kidneys, for years or decades.

The toxicity of metals most commonly involves the brain and the kidney, but other manifestations occur, and some metals, such as arsenic, are clearly capable of causing cancer. An individual with metals toxicity, even if high dose and acute, typically has very general symptoms, such as weakness or headache. This makes the diagnosis of metals toxicity in a clinical setting very difficult unless a clinician has the knowledge and training to suspect the diagnosis and is able to order the correct diagnostic test. Chronic exposure to metals at a high enough level to cause chronic toxicity effects (such as hypertension in individuals exposed to lead and renal toxicity in individuals exposed to cadmium) can also occur in individuals who have no symptoms.

Much about metals toxicity, such as the genetic factors that may render some individuals especially vulnerable to damage, remains a subject of intense investigation. It is possible that low-level metals exposure contributes much more toward the causation of chronic disease and impaired functioning than previously thought.

This chapter focuses on exposure to the four "heavy" metals on the ATSDR list mentioned above—lead, mercury, arsenic, and cadmium—because they are arguably the most important metal toxins from a global, as well as a U.S., perspective. Some additional remarks are also made regarding a few other metals of concern. (Exposure to arsenic and lead in drinking water is covered by John Balbus in chapter 3.)

Lead

Lead is the most widely used and the most carefully studied of the heavy metals. The lessons learned from lead about exposure, toxicity, and industrial practice are now being applied in the study of the other metals.

For centuries, lead has been mined and used in industry and in household products. Modern industrialization, with the introduction of lead in mass-produced plumbing, solder used in food cans, paint, ceramic ware, and countless other products, resulted in a marked rise in population exposures in the twentieth century.

The dominant source of worldwide dispersion of lead into the environment (and into people) for the past 50 years has clearly been the use of lead organic compounds as antiknock motor vehicle fuel additives. Since leaded gasoline was introduced in 1923, its combustion and resulting contamination of the atmosphere has increased background levels everywhere, including the ice cap covering Northern Greenland (figure 4.1), where there is no industry and few cars and people.[1] Although a worldwide phaseout of leaded gasoline is in progress (see http://www.earthsummit-watch.org/gasoline.html for details), it is still being used all over the world.

The current annual worldwide production of lead is approximately 5.4 million tons and continues to rise. Sixty percent of lead is used for the manufacturing of batteries (automobile batteries, in particular), while the remainder is used in the production of pigments, glazes, solder, plastics, cable sheathing, ammunition, weights, gasoline additives, and a variety of

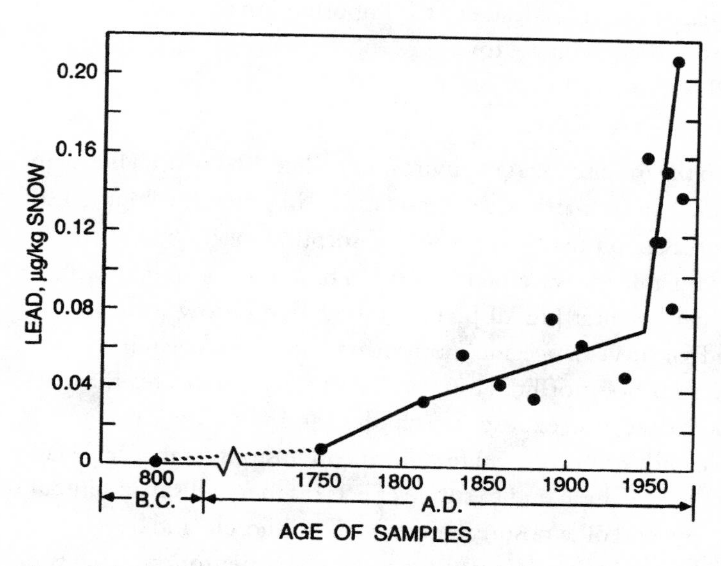

Figure 4.1 Levels of lead in the ice cap covering Northern Greenland (reference 1)

other products. Such industries continue to pose a significant risk to workers, as well as to surrounding communities.

In response to these risks, many developed countries over the last 25 years have implemented regulatory action that has effectively decreased actual exposures to the general population. However, exposures remain high or are increasing in many developing countries through a rapid increase in vehicles combusting leaded gasoline and in polluting industries (some of which have been "exported" by corporations in developed countries seeking relief from regulations). Moreover, some segments of the population in developed countries (such as the United States) remain at high risk of exposure because of the persistence of lead paint, lead plumbing, and lead-contaminated soil and dust, particularly in areas of old urban housing.

A number of factors can modify the impact of lead exposures. For example, water with a lower pH (such as drinking water stemming from the collection of untreated "acid rain") will leach more lead out of plumbing connected by lead solder than more alkaline water will.[2] Lead from soil tends to concentrate in root vegetables (e.g., onions)[3] and leafy green vegetables (e.g., spinach). Individuals will absorb more lead in their food if their diets are deficient in calcium, iron, or zinc.[4] Other more unusual sources of lead exposure also continue to be sporadically found, such as improperly glazed ceramics, lead crystal, imported candies, certain herbal folk remedies, and vinyl plastic toys.

Toxicity

Lead has been the intense focus of environmental health research for many decades. Studies in humans were greatly assisted by the development of methods (such as graphite furnace atomic absorption spectroscopy) for the accurate and reliable measurement of lead in blood (measured in units of micrograms per deciliter [μg/dL]), a technique that is now widely available and used for surveillance and monitoring, as well as research.

The general body of literature on lead toxicity indicates that, depending on the dose, lead exposure in children and adults can cause a wide spectrum of health problems, ranging from convulsions, coma, renal failure, and death at the high end to subtle effects on metabolism and intelligence at the low end of exposures[5] (figure 4.2). Children (and developing fetuses) appear to be particularly vulnerable to the neurotoxic effects of

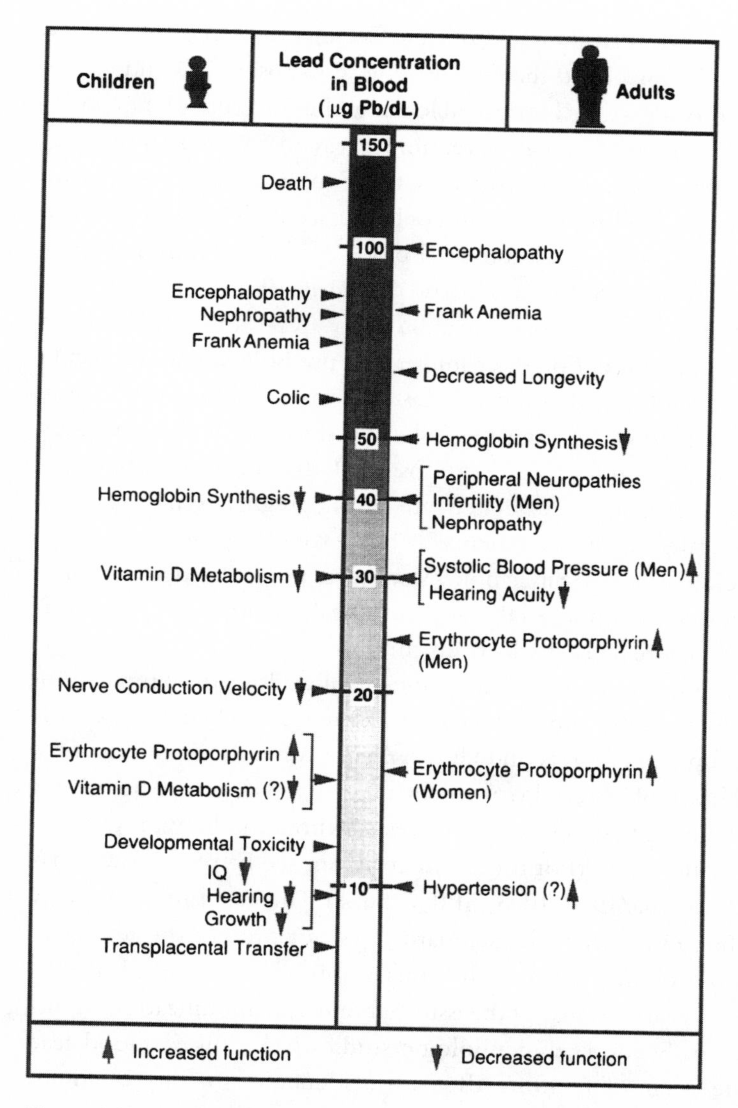

Figure 4.2 Health effects associated with various levels of lead in the blood of children and adults (reference 2)

lead. A plethora of well-designed prospective epidemiologic studies has convincingly demonstrated that low-level lead exposure in children less than five years of age (with blood lead levels in the 5–25 μg/dL range) results in deficits in intellectual development, as manifested by lost IQ points.[6] As a result, in the United States, the Centers for Disease Control (CDC) lowered the allowable amount of lead in a child's blood from 25 to 10 μg/dL and recommended universal blood-lead screening of all children between the ages of six months and five years.[7] (For more details, see http://www.cdc.gov/nceh/lead/lead.htm.)

However, a number of issues still remain unresolved with respect to lead toxicity in children. Among the most important is the risk to the fetus posed by mobilization of long-lived skeletal stores of lead in pregnant women.[8] Research has clearly demonstrated that maternal bone-lead stores are mobilized at an accelerated rate during pregnancy and lactation[9] and are associated with decrements in birth weight, growth rate, and mental development.[10–12] Since bone-lead stores persist for decades,[13] it is possible that lead can remain a threat to fetal health many years after environmental exposure had actually been curtailed.

In contrast to children, adults are generally allowed by regulations to be exposed to higher amounts of lead. In the United States, for example, the Occupational Safety and Health Administration requires that the blood-lead levels of exposed workers be maintained below 40 μg/dL as a way of preventing toxic effects to nerves, the brain, kidneys, reproductive organs, and heart. (For more information, see http://www.osha-slc.gov/OshStd_data/1910_1025_APP_C.html.) This standard is probably outdated, however. First, the standard does not protect the fetuses of women who become pregnant while on the job (or even if they leave the job for several years because of the issue of bone-lead mobilization, as discussed above). Second, epidemiologic studies have linked blood-lead levels in the range of 7–40 μg/dL with evidence of toxicity in adults, such as neurobehavioral decrements[14] and renal impairments.[15] Third, recent studies using a newly developed technique, K-x-ray fluorescence, to directly measure bone-lead levels (as opposed to blood-lead levels) have provided evidence demonstrating that cumulative lead exposure in individuals with blood-lead levels well below 40 μg/dL is a major risk factor for the development of hypertension,[16–18] cardiac conduction delays,[19] and cognitive impairments.[20,21]

Finally, even as research progresses to delineate the full toxicologic implications of lead exposure, investigations at the interface of genetics and environmental health are beginning to uncover subgroups of individuals who may be particularly susceptible to the toxicity of lead.[22]

Mercury

Mercury comes in a number of different chemical forms. Metallic mercury (Hg) is used in thermometers, dental amalgams, and some batteries.

Exposure

In its pure form, metallic mercury is a liquid. Contrary to popular opinion, it is not hazardous if ingested (because it is not significantly absorbed in this form). However, if left standing, or even worse, if aerosolized—for example, through attempted vacuum cleaning—metallic mercury will volatilize into a vapor that is well absorbed by the lungs. Mercurous and mercuric mercury (Hg^{+} and Hg^{2+}, respectively) are encountered in some chemical, metal-processing, electrical-equipment, automotive, and building industries and in medical and dental services. Mercurous and mercuric mercury form inorganic and organic compounds with other chemicals that can be readily absorbed through ingestion. All three forms of mercury are toxic to various degrees.

From a global perspective, mercury has been increasing in importance as a widespread contaminant. About half of the National Priority List toxic waste sites in the United States contain mercury. Mercury dispersion through atmospheric deposition has increased markedly through waste incineration; ironically, the medical industry is one of the largest contributors to mercury pollution in this fashion.[23] Some countries, such as Brazil, have seen widespread mercury contamination (and resultant health effects) through a combination of its indiscriminate use in gold mining and deforestation.[24] When deposited in soil, organic mercury compounds are slowly broken down into inorganic compounds; conversely, inorganic mercury can be converted by microorganisms in soil and water into the organic compound methyl mercury, which is then bioconcentrated up the food chain. Fish, particularly tuna, king mackerel, and swordfish, can concentrate methyl mercury at high levels. Their consumption, though popular, should be limited in frequency (e.g., to less than twice per week) and entirely avoided by women during pregnancy.

Toxicity

High levels of mercury exposure that occur through, for example, inhalation of mercury vapors generated by thermal volatilization can lead to life-threatening injuries to the lungs and neurological system. At lower but more chronic levels of exposure, a typical constellation of findings arises, termed *erethism*—with tremor of the hands, excitability, memory loss, insomnia, timidity, and sometimes delirium—that was once commonly seen in workers exposed to mercury in the felt-hat industry ("mad as a hatter"). Even relatively modest levels of occupational mercury exposure, as experienced, for example, by dentists, have been associated with measurable declines in performance on neurobehavioral tests of motor speed, visual scanning, verbal and visual memory, and visuomotor coordination.[25] Contrary to some opinions expressed in the popular media, however, evidence from well-conducted studies is lacking that the small amount of mercury released from dental amalgams during chewing is capable of causing significant illnesses, such as mutliple sclerosis, systemic lupus, or chronic fatigue syndrome.[26] Special note should be taken of dimethylmercury—a "supertoxic, superdangerous" compound that can penetrate through latex gloves, as well as skin. This compound does not occur naturally; rather, it is synthesized for use as a standard in nuclear magnetic resonance spectroscopy of mercury compounds and for heavy-atom labeling of biomolecules for x-ray structure determination. Following a recent fatality in which exposure to only a few drops spilled on a latex-gloved hand resulted in systemic absorption of the compound, central nervous system degeneration, and the death [27] of a prominent environmental chemist, it is likely that the continued use of dimethylmercury for even these few specialized applications will be curtailed.

Of greatest concern on a global scale is the sensitivity of the fetal and infant nervous system to low-level mercury toxicity. Mothers exposed to mercury in the 1955 disaster in Minamata Bay, Japan, gave birth to infants with mental retardation, retention of primitive reflexes, cerebellar symptoms, and other abnormalities. Recent research in the Faroe Islands has demonstrated that, even at much lower levels, mercury exposure on the part of pregnant women through dietary intake of fish and whale meat, an important regional food staple, is associated with decrements in motor function, language, memory, and neural transmission in their offspring.[28,29] Organic mercury, the form of mercury bioconcentrated in fish and whale meat, readily crosses the placenta and appears in breast milk.

Arsenic

Arsenic is a toxic and carcinogenic heavy metal, and human exposures are common. Scientists and regulators are debating safe exposure standards.

Exposure

Significant exposure to arsenic occurs through both anthropogenic and natural sources. Occupational and community exposures to arsenic from the activities of humans occur through the smelting industry, the use of gallium arsenide in the microelectronics industry, and the use of arsenic in common products such as wood preservatives, pesticides, herbicides, fungicides, and paints. Widespread dispersion of arsenic is a by-product of the combustion of fossil fuels in which arsenic is a common contaminant. Arsenic continues to be found in some folk remedies.

In some areas of the world, arsenic is also a natural contaminant of wells. Deep-water wells in parts of Taiwan and Chile are now well known to be contaminated with arsenic, giving rise to chronic manifestations of toxicity discussed below.[30,31] Water from relatively shallow tube wells that were placed in areas of Bangladesh, West Bengal, and other parts of the subcontinent has also been found to be heavily contaminated with arsenic[32]—a particularly unfortunate revelation given the dire need for potable drinking water in the populations living in these areas. Water in some parts of the United States, such as areas of the Southwest, also carry a significant risk of arsenic contamination. (See chapter 3.)

Toxicity

The toxicity of an arsenic-containing compound depends on its valence state (zero-valent, trivalent, or pentavalent), its form (inorganic or organic), and factors that modify its absorption and elimination. Inorganic arsenic is generally more toxic than organic arsenic, and trivalent arsenite is more toxic than pentavalent and zero-valent arsenic. These nuances are important. For example, testing biological samples for arsenic in an individual with suspected toxicity must be done more than 48 hours after the individual abstains from eating seafood; otherwise, the test may be confounded by the presence of arsenobentaine, a relatively harmless form of arsenic contained in fish at high levels of concentration.

Once absorbed into the body, arsenic undergoes some accumulation in soft-tissue organs such as the liver, spleen, kidneys, and lungs, but the major long-term storage site for arsenic is keratin-rich tissues, such as skin, hair, and nails—making the measurement of arsenic in these biological specimens useful for estimating total arsenic burden and long-term exposure under certain circumstances.

Acute arsenic poisoning is infamous for its lethality, which stems from arsenic's destruction of the integrity of blood vessels and gastrointestinal tissue and its effect on the heart and brain. Chronic exposure to lower levels of arsenic results in somewhat unusual patterns of skin hyperpigmentation, peripheral nerve damage manifesting as numbness, tingling, and weakness in the hands and feet, diabetes, and blood-vessel damage resulting in a gangrenous condition affecting the extremities.[33] Chronic arsenic exposure also causes a markedly elevated risk for developing a number of cancers, most notably skin cancer, cancers of the liver (angiosarcoma), lung, bladder, and possibly the kidneys and colon. The dose necessary to increase the risk for cancer has become the focus of particularly intense scrutiny in the United States because of proposed efforts to lower standards governing general-population exposures to arsenic. Among environmental scientists studying this problem, the most common view is that the current standard for the allowable amount of arsenic in U.S. drinking water—50 μg/liter—is probably not adequate to sufficiently safeguard the general population from arsenic's cancer risk.[34]

Cadmium

Acute cadmium toxicity is relatively uncommon, but exposure to this metal causes distinctive clinical syndromes.

Exposure

Cadmium exposure is encountered in industries dealing with pigment, metal plating, some plastics, and batteries. Cadmium pollution (e.g., the emissions of a cadmium smelter or industry and the introduction of cadmium into sewage sludge, fertilizers, and groundwater) can result in significant human exposure to cadmium through the ingestion of contaminated foodstuffs, especially grains, cereals, and leafy vegetables. Airborne cadmium exposure is also a risk posed by the incineration of municipal waste containing plastics and nickel-cadmium batteries. Cigarette smoking constitutes an additional major source of cadmium exposure.

Toxicity

The health implications of cadmium exposure are exacerbated by the relative inability of human beings to excrete cadmium. (It is excreted but then reabsorbed by the kidneys.) Acute high-dose exposures can cause severe respiratory irritation. Occupational levels of cadmium exposure are a risk factor for chronic lung disease (through airborne exposure) and testicular degeneration[35] and are still under investigation as a risk factor for prostate cancer.[36] Lower levels of exposure are mainly of concern with respect to toxicity to the kidneys. Cadmium damages a specific structure of the functional unit of the kidney (the proximal tubules of each nephron) in a way that is first manifested by leakage of low-molecular-weight proteins and essential ions, such as calcium, into urine, with progression over time to kidney failure.[37] This effect tends to be irreversible,[38] and recent research suggests that the risk exists at lower levels of exposure than previously thought.[39,40] Even without causing kidney failure, however, cadmium's effect on the kidneys can have metabolic effects with pathological consequences. In particular, the loss of calcium caused by cadmium's effect on the kidneys can be severe enough to lead to weakening of the bones. "Itai-itai" disease, an epidemic of bone fractures in Japan from gross cadmium contamination of rice stocks, has recently been shown to happen in more subtle fashion among a general community living in an area of relatively modest cadmium contamination.[41] Increased cadmium burden in this population was found to be predictive of an increased risk of bone fractures in women, as well as decreased bone density and height loss (presumably from the demineralization and compression of vertebrae) in both sexes.

Other Metals of Concern

Manganese has become a metal of global concern because of the introduction of methylcyclopentadienyl manganese tricarbonyl (MMT) as a gasoline additive. Proponents of the use of MMT have claimed that the known link between occupational manganese exposure and the development of a Parkinson's disease–like syndrome of tremor, postural instability, gait disorder, and cognitive disorder has no implications for the relatively low levels of manganese exposure that would ensue from its use in gasoline. However, this argument is reminiscent of the rationale given for adding lead to gasoline, and what little research exists from which one can infer the toxicity potential of manganese at low levels of exposure is not particularly comforting.[42]

Aluminum contributes to the brain dysfunction of patients with severe kidney disease who are undergoing dialysis. Dialysis equipment is made of aluminum, and the metal leaches into the dialysis fluid and hence directly into the patient's circulation. High levels of aluminum have been found in neurofibrillary tangles (characteristic brain lesions in patients with Alzheimer's disease), as well as in the drinking water and soil of areas with an unusually high incidence of Alzheimer's disease. Nevertheless, the experimental and epidemiologic evidence for a causal link between aluminum exposure and Alzheimer's disease is, overall, relatively weak. More research is needed on this topic, since general-population exposures to aluminum are increasing through the use of aluminum cookware, aluminum-containing deodorants, and other products.

Chromium, in its hexavalent form, which is the most toxic species of chromium, is used extensively in some industries such as leather processing. As a result, chromium has become a major factory runoff pollutant that is beginning to become a global trend. The toxicity of chromium stems from its tendency to be corrosive and to cause allergic reactions. Chromium is a carcinogen, particularly of the lung through inhalation.

Prescriptions to Reduce Human Exposure to Heavy Metals

Lead has been emphasized in this review because, in many ways, it is the archetype metal toxin in terms of its many sources and pathways for exposure, global distribution, ability to accumulate in the body, and impact on human health even at low levels of exposure. Unfortunately, even though lead is arguable the most-studied toxin in all of environmental health, a host of unresolved issues remain, such as whether and to what degree certain subpopulations are especially susceptible to lead toxicity (through genetic variations), to what degree lead's effect on mental development in children is reversible, and whether lead plays a role in the development of chronic neurodegenerative diseases such as Alzheimer's disease and Parkinson's disease. The number of unresolved issues surrounding the other toxic metals of concern is even larger. Nevertheless, even without additional basic research on the toxicity of metals, enough is known to warrant close attention to and efforts at controlling metals contamination.

Prescription 1: Accelerate and complete the global phaseout of leaded gasoline. Some countries, like Italy, Greece, Spain, Australia, and Chile, lag behind in their switch to unleaded gasoline; others, such as South Africa, Nigeria, the Confederation of Independent States, Israel, and Malaysia, have yet to even commit to this process. (See http://www.earthsummitwatch.org/gasoline.html for more details.) Delays are inadvisable, given the accumulation of knowledge on low-level lead toxicity reviewed above.

Prescription 2: Begin an effort to monitor levels and trends in metals pollution worldwide. Currently, no agency or database exists from which one can gain a clear sense of levels and trends in pollution by metals. Some monitoring occurs in the use and distribution of exposure vectors—such as worldwide trends in the use of leaded gasoline (see above)—but the sources of exposure to metals are diverse and numerous, making it difficult to infer the impact of such exposure vector trends on exposure at the individual level. A clearer picture would emerge if samples were regularly taken, using standardized methods for sampling and analysis, of metal levels in air, drinking water, and market-basket surveys of food.

Prescription 3: Establish population-based biomonitoring for selected metals. Exposure to some metals, such as lead, can be best monitored by following trends in blood or other media (such as urine). Many such surveys appear regularly in the literature, but they are haphazard and depend on the specific (and changing) interests of their sponsors. The principal challenges to establishing such a system are defining test populations that are fairly representative of national/regional trends, standardizing technical protocols for analysis of samples, and organizing and funding such an effort.

Prescription 4: Educate governments, scientists, and the general public about the toxicity of metals. The information above, while in the public domain, is not easily accessed or grasped by average citizens and policymakers and is rarely taught in schools of medicine, nursing, or other allied health professions.

Prescription 5: Declare a moratorium on the production, distribution, and use of products likely to significantly increase global exposure to toxic metals. The most obvious target for this prescription is the example of methylcyclopentadienyl manganese tricarbonyl (MMT), the newly created additive for gasoline. It makes little sense to globally disperse a new metal for which the toxicological implications are not fully known (or, in the case of MMT, known to be severe at high levels of exposure).

Prescription 6: Continue basic research into the impact of metals on human health. As described in various places earlier in this chapter, numerous unresolved issues remain regarding the impact of metals on human health. Current research priorities (and funding) in medical research is heavily skewed toward treatment of chronic diseases such as cancer, heart disease, neurological diseases, and osteoporosis. Attention should be increased toward research on prevention, the most fundamental aspect of which is investigating the causes and relative importance of various risk factors for the diseases of interest.

References

1. US Environmental Protection Agency. *Air quality criteria for lead.* Research Triangle Park, NC: Environmental Criteria and Assessment Office, 1986 Jun. EPA–600/8–83–028.

2. Moore MR. Influence of acid rain upon water plumbosolvency. *Environ Health Perspect* 1985; 63:121–126.

3. Ward NI, Savage JM Metal dispersion and transportational activities using food crops as biomonitors. *Sci Total Environ* 1994; 146:309–319.

4. Mahaffey KR. Environmental lead toxicity: Nutrition as a component of intervention. *Environ Health Perspect* 1990; 89:75–78.

5. US Agency for Toxic Substances and Disease Registry. *Lead. Toxicological profiles.* Atlanta: Centers for Disease Control and Prevention, 1999. PB/99/166704.

6. Banks EC, Ferretti LE, Shucard DW. Effects of low-level lead exposure on cognitive function in children: A review of behavioral, neuropsychological, and biological evidence. *Neurotoxicology* 1997; 18: 237–281.

7. Centers for Disease Control. *Preventing lead poisoning in young children: A statement by the U.S. Centers for Disease Control—October 1991.* US Department of Health and Human Services, 1991.

8. Silbergeld EK. Lead in bone: Implications for toxicology during pregnancy and lactation. *Environ Health Perspect* 1991; 91:63–70.

9. Gulson BL, Jameson CW, Mahaffey KR, Mizon KJ, Korsch MJ, Vimpani G. Pregnancy increases mobilization of lead from maternal skeleton. *J Lab Clin Med* 1997; 30:51–62.

10. Gonzalez-Cossio T, Peterson KE, Sanin L, Fishbein SE, Palazuelos E, Aro A, Hernández-Avila M, Hu H. Decrease in birth weight in relation to maternal bone-lead burden. *Pediatrics* 1997; 100:856–862.

11. Sanin LH, Gonzalez-Cossio T, Romieu I, Peterson KE, Ruiz S, Hernandez-Avila M, et al. Effect of maternal lead burden on infant weight gain at one month of age among breastfed infants. *Pediatrics* 2001; 107:1016–1023.

12. Gomaa A, Hu H, Bellinger D, Schwartz J, Schnaas L, Gonzalez-Cossio T, et al. Maternal bone lead as an independent risk factor for fetal neurotoxicity: A prospective study. *Pediatrics* (in press).

13. Hu H, Rabinowitz M, Smith D. Bone lead as a biological marker in epidemiologic studies of chronic toxicity: Conceptual paradigms. *Environ Health Perspect* 1998; 106:1–7.

14. Schwartz BS, Lee BK, Lee GS, Stewart WF, Lee SS, Hwang KY, et al. Associations of blood lead, dimercaptosuccinic acid-chelatable lead, and tibia lead with neurobehavioral test scores in South Korean lead workers. *Am J Epidemiol* 2001; 153:453–464.

15. Kim R, Rotnitzky A, Sparrow D, Weiss ST, Wager C, Hu H. A longitudinal study of low-level lead exposure and impairment of renal function: The Normative Aging Study. *JAMA* 1996; 275:1177–1181.

16. Hu H, Aro A, Payton M, Korrick S, Sparrow D, Weiss ST, Rotnitzky A. The relationship of bone and blood lead to hypertension: The Normative Aging Study. *JAMA* 1996; 275:1171–1176.

17. Korrick SA, Hunter DJ, Rotnitzky A, Hu H, Speizer FE. Lead and hypertension in a sample of middle-aged women. *Am J Public Health* 1999; 89:330–335.

18. Cheng Y, Schwartz J, Sparrow D, Aro A, Weiss ST, Hu H. A prospective study of bone lead level and hypertension: The Normative Aging Study. *Am J Epidemiol* 2001; 153:164–171.

19. Cheng Y, Schwartz J, Vokonas P, Weiss ST, Aro A, Hu H. Electrocardiographic conduction disturbances in association with low level lead exposure: The Normative Aging Study. *Am J Cardiol* 1998; 82:594–599.

20. Stewart WF, Schwartz BS, Simon D, Bolla KI, Todd AC, Links J. Neurobehavioral function and tibial and chelatable lead levels in 543 former organolead workers. *Neurology* 1999; 52:1610–1617.

21. Payton M, Riggs KM, Spiro A, Weiss ST, Hu H. Relations of bone and blood lead to cognitive function: The VA Normative Aging Study. *Neurotox Teratol* 1998; 20:19–27.

22. Onalaja AO, Claudio L. Genetic susceptibility to lead poisoning. *Environ Health Perspec* 2000; 108 Suppl 1:23–38.

23. Harvie, J. Eliminating mercury use in hospital laboratories: A step toward zero discharge. *Public Health Rep* 1999; 114:353–358.

24. Dolbec J, Mergler D, Sousa Passos CJ, Sousa de Morais S, Lebel J. Methylmercury exposure affects motor performance of a riverine population of the Tapajos river, Brazilian Amazon. *Int Arch Occup Environ Health* 2000; 73:195–203.

25. Bittner AC Jr, Echeverria D, Woods JS, Aposhian HV, Naleway C, Martin MD, et al. Behavioral effects of low-level exposure to Hg° among dental professionals: A cross-study evaluation of psychomotor effects. *Neurotoxicol Teratol* 1998; 20:429–349.

26. Grandjean P, Guldager B, Larsen IB, Jorgensen PJ, Homstrup P. Placebo response in environmental disease. Chelation therapy of patients with symptoms attributed to amalgam fillings. *J Occup Environ Med* 1997; 39:707–714.

27. Nierenberg DW, Nordgren RE, Change MB, Siegler RW, Blayney MB, Hochberg F, et al. Delayed cerebellar disease and death after accidental exposure to dimethylmercury. *New Engl J Med* 1998; 338:1672–1676.

28. Grandjean P, Weihe P, White RF, Debes F. Cognitive performance of children prenatally exposed to "safe" levels of methylmercury. *Environ Res* 1998; 77:165–172.

29. Murata K, Weihe P, Araki S, Budtz-Jorgensen E, Grandjean P. Evoked potentials in Faroese children prenatally exposed to methylmercury. *Neurotoxicol Teratol* 1999; 21:471–472.

30. Tsai SM, Wang TN, Ko YC. Morality for certain diseases in areas with high levels of arsenic in drinking water. *Arch Environ Health* 1999; 43:186–193.

31. Ferreccio C, Gonzalez C, Milosavjlevic V, Marshall G, Sancha AM, Smith AH. Lung cancer and arsenic concentrations in drinking water in Chile. *Epidemiology*. 2000; 11:673–679.

32. Chowdhury UK, Biswas BK, Chowdhury TR, Samanta G, Mandal BK, Basu GC, et al. Groundwater arsenic contamination in Bangladesh and West Bengal, India. *Environ Health Perspect* 2000; 108:393–397.

33. Col M, Col C, Soran A, Sayli BS, Ozturk S. Arsenic-related Bowen's disease, palmar keratosis, and skin cancer. *Environ Health Perspect* 1999; 107:687–689.

34. Morales KH, Ryan L, Kuo TL, Wu MM, Chen CJ. Risk of internal cancers from arsenic in drinking water. *Environ Health Perspect* 2000; 108:655–661.

35. Benoff S, Jacob A, Hurley IR Male infertility and environmental exposure to lead and cadmium. *Hum Reprod Update* 2000; 6:107–121.

36. Ye J, Wang S, Barger M, Castranova V, Shi X. Activation of androgen response element by cadmium: A potential mechanism for a carcinogenic effect of cadmium in the prostate. *J Environ Pathol Toxicol Oncol* 2000; 19:275–280.

37. Satarug S, Haswell-Elkins MR, Moore MR. Safe levels of cadmium intake to prevent renal toxicity in human subjects. *Br J Nutr* 2000; 84:791–802.

38. Roels HA, Van Asche FJ, Oversteys M, De Groof M, Lauwerys RR, Lison D. Reversibility of microproteinuria in cadmium workers with incipient tubular dysfunction after reduction of exposure. *Am J Ind Med* 1997; 31:645–652.

39. Suwazono Y, Kobayashi E, Okubo Y, Nogawa K, Kido T, Nakagawa H. Renal effects of cadmium exposure in cadmium nonpolluted areas in Japan. *Environ Res* 2000; 84:44–55.

40. Jarup L, Hellstrom L, Alfven T, Carlsson MD, Grubb A, Persson B, et al. Low level exposure to cadmium and early kidney damage: The OSCAR study. *Occup Environ Med* 2000; 57:668–672.

41. Staessen JA, Roels HA, Emelianov D, Kuznetsova T, Thijs L, Vangronsveld J, et al. Environmental exposure to cadmium, forearm bone density, and risk of fractures: Prospective population study. Public Health and Environmental Exposure to Cadmium (PheeCad) Study Group. *Lancet* 1999; 353:1140–1144.

42. Lyzincki JM, Karlan MS, Khan MK. Manganese in gasoline. Council on Scientific Affairs, American Medical Association. *J Occup Environ Med* 1999; 41:140–143.

Population, Consumption, and Human Health

J. Joseph Speidel

5

Strong evidence indicates that the growth of the world population poses serious threats to human health, socioeconomic development, and the environment.[1,2] In 1992, the Union of Concerned Scientists issued a World Scientists' Warning to Humanity, signed by 1,600 prominent scientists, that called attention to threats to life-sustaining natural resources.[3] In 1993, a Population Summit of 58 of the world's scientific academies voiced concern about the intertwined problems of rapid population growth, wasteful resource consumption, environmental degradation, and poverty.[4] These reports share the view that, without stabilization of both population and consumption, good health for many people will remain elusive, developing countries will find it impossible to escape poverty, and environmental degradation will worsen.

Population Growth

It took only 12 years for the world population to grow from 5 billion to 6 billion in the year 2000. This is the shortest time ever to add 1 billion people—a number equivalent to the population of India or the combined population of the United States and Europe.

Over the 17 centuries ending in 1800, the world population grew slowly, from an estimated 250 million to about 1 billion. Over the past two centuries, and especially after 1950, declining death rates brought about rapid growth. By 1950, the world population had reached 2.5 billion, the world total fertility rate (TFR: the mean lifetime number of children borne by each woman) was 5.3, and the population was growing by about 40 million per year.[5,6] Since 1950, the world TFR has declined to 2.8; however, continued declines in death rates and the growth of the population to 6.1 billion together have led to the doubling of annual growth in the world population to 82 million.[7]

Over the past 200 years, Western nations have made a gradual demographic move from high to low birth rates and death rates. These countries are now growing by only 0.1 percent annually.[7] Over the past 50 years, public health measures and improved nutrition in developing countries have rapidly lowered death rates. Although use of family planning in these countries has increased substantially (from about 10 percent of couples to over 50 percent), greater use of contraception is hampered by poverty, lack of education, and inadequate access to family planning information and services.[8,9] As a result, declines in birth rates in developing countries have been uneven and have usually lagged behind declines in death rates. Therefore, growth rates have remained relatively high.[5–7,9] Currently, about 97 percent of population growth is occurring in developing countries, which, between 1987 and 1999, grew by 1 billion people.[7,10]

The United Nations (UN) recently presented three demographic projections for the next 100 years that, although rapid declines in fertility are expected, still see substantial increases in the world population.[11,12] Because population projections are extremely sensitive to fertility rates, the accuracy of long-range projections is uncertain.[12,13]

The UN's medium-fertility projection suggests a decline from the current world TFR of 2.8 to 2.1 by 2050, with a resulting population size of 9.3 billion that will continue to grow slowly to about 10.4 billion by 2100.[11,12] The U.S. Census Bureau projections of growth and total world population size through 2050 are shown in figure 5.1.

The UN's high-fertility model assumes that in countries with high fertility rates (TFRs above replacement level), the TFRs will stabilize at 2.6 and that in countries with low fertility rates (TFRs currently below 2.1), the TFRs will increase and stabilize at between 2.1 and 2.3. This model projects that the world population will continue to grow rapidly, reaching 10.9 billion by 2050.[11] If fertility remains constant at year-2000 levels, world population would reach 13 billion by 2050.[11]

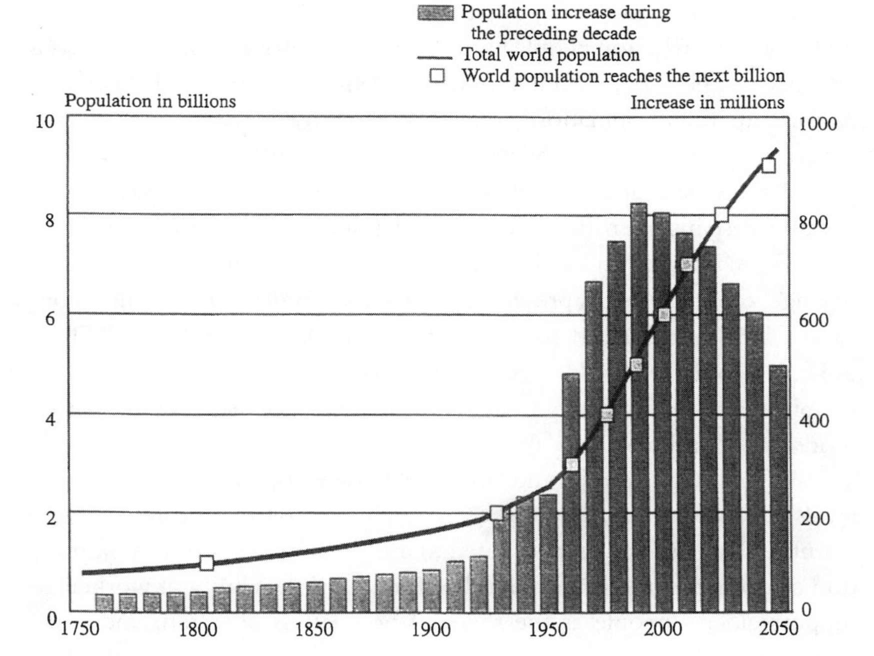

Figure 5.1 World population from 1750 to 2050 (black line), and population increases by decade (gray bars) (reference 14)

The UN's low-fertility model, which assumes worldwide TFRs of 1.7 by 2050, projects an initial increase in population size, followed by a slow decline to 7.8 billion in 2050.[11] In part because support for international family planning programs remains inadequate, the low-fertility projection seems unlikely. Without greatly strengthened efforts to provide family planning services, even the medium projection is in doubt.[15–17]

The fertility of the world's developed countries is now at a TFR of 1.6,[7] so low that gradual population decline can be expected in most of these countries.[11] The fertility rates of some 25 developing countries, including those in East Asia and China, are already at or are likely to soon decline to or go below replacement level. Some 44 percent of the world's people now live in low-fertility countries (20 percent in developed countries, 20 percent in China, and 4 percent in other developing countries), and UN population projections suggest that more developing countries will reach fertility levels below replacement level.[11,12]

However, high fertility persists in much of the world. The current TFR for the 3.6 billion people living in poor countries outside of China is estimated to be 3.7 and their annual population growth rate 1.9 percent. At this rate, their population would double in just 36 years. Despite projected declines in fertility, the number of annual births worldwide are expected to remain at over 130 million for the next 25 years.[14] This is because past high fertility rates in most poor countries have left these countries with large and still increasing numbers of women of reproductive age; their number is projected to increase from 1.2 to 1.7 billion between 1998 and 2025.[14] In China, for example, although the TFR is estimated to be below the replacement level of 2.1, the large number of couples of reproductive age have kept China's population growing by more than 11 million annually.[7,11]

Even though Europe and Japan are densely populated and have high levels of consumption, the prospect of gradual population decline in these countries has raised concerns related to immigration, the ethnic composition of countries, the size of the labor force, and the ability of workers to support elderly people as the share of population over retirement age increases.

The United States and Canada are, to some extent, exceptions among Western nations. Because of high levels of immigration, their populations are still growing relatively rapidly. The U.S. population is projected to increase from 281 million to about 400 million (or even 500 million) by 2050.[18,19] The arrival of about 1.3 million people per year (800,000 legal and 500,000 illegal immigrants) and the high fertility rate among the 30 million foreign-born residents is fueling this growth.[18,19] Similar projections for Canada suggest an increase from the current 31 million to over 42 million in 2050, with over half of this growth the result of immigration.[20,21] Considering the high level of individual consumption in the United States and Canada, a 42 percent increase in population would have profound implications for land, air, and water resources.

Interactions between Population, Consumption, the Environment, and Health

Some 10,000 years ago, when only about 5 million people inhabited the earth, few biological systems were seriously damaged by human activity.

Today, however, the world faces an environmental dilemma. Current demands are depleting many of the earth's natural resources and ecological services.[22–24] Within the next 50 years, it is likely that those life-supporting systems will somehow have to accommodate 3 billion more people, as well as support desperately needed advances in living standards for those in poverty, particularly the 3 billion people now living on about $2 a day.[11,25,26]

The impact of humans on their environment is related to population size, per capita consumption, and the environmental damage caused by the technology used to produce what is consumed. The exploitation of technology and the high consumption pattern of people in Japan, Europe, the United States, and Canada have a greater adverse impact per capita on the world's environment than that of a subsistence farmer in Bangladesh, for example. Although they represent 20 percent of the world's population, the 1.2 billion people living in developed countries consume an estimated 67 percent of all resources and generate 75 percent of all waste and pollution.[22–24]

Between 1950 and 1997, the world's population doubled and the global economy expanded sixfold, from $5 trillion to $29 trillion of annual output.[27] A further modest 2 percent annual growth in incomes and consumption per capita worldwide could result in a doubling of consumption every 35 years, or about an eightfold increase by the year 2100. This increased consumption per capita, on top of a projected population increase of 1.73 times, from 6 to 10.4 billion,[11,12] would require economic production to increase fourteenfold. Achieving this without substantial degradation of important ecosystems presents a daunting challenge.

Many important interactions exist between population growth, consumption, environmental degradation, and health. Human activity has already transformed an estimated 50 percent of the earth's surface, with 10 percent of land degraded from forest or rangeland into desert. The productive capacity of 25 percent of all agricultural lands, an area equal to the size of India and China combined, has already been degraded.[24,28] Unproductive land and food scarcity currently contribute to malnutrition among 1 billion people, with infants and children suffering the most serious health consequences.[29,30]

Projected population growth in Africa, South Asia, and other developing areas, together with declining availability of water from aquifers, threatens the food security of more than 1 billion people in developing countries. Studies have indicated that depletion of aquifers threatens India with a 25 percent decline in grain production, at a time when over half of the country's children are malnourished and the population is projected to increase by some 500 million over the next 50 years.[31,32] Other large countries where rapid population growth and declining cropland per person threaten food security include Nigeria and Pakistan. If Nigeria's population increases from the current 111 million to a projected 244 million in 2050, grainland per capita will decline from 0.15 to 0.07 hectares. The corresponding projection for Pakistan is an increase in population from 146 million to 345 million and a shrinkage of grainland per person from 0.08 to 0.03 hectares.[33] Countries with these levels of grainland typically import over half of their grain, an expensive and perhaps impossible prospect for these impoverished countries.[31,32]

Water scarcity also impairs health, as freshwater supplies for human use become polluted with toxic materials and pathogens. Proper treatment of human waste is currently not available for about 2 billion people, and 1.3 billion people are at risk of waterborne diseases because they lack access to pure drinking water.[22,24,34]

There is growing evidence that global warming is occurring, increasing the prospect of flooded coastal areas and cities, disruptions of agriculture, increasingly severe storm damage,[30–35,36] and significant extension of the range of insects and other vectors of disease.[37] Environmental degradation, declining food security, and uncontrolled epidemics of communicable diseases have slowed, and even reversed, the demographic transition to low death rates in some poor countries. In contrast to developed countries, where cardiovascular diseases and cancer are the leading causes of death, infectious diseases cause 45 percent of all deaths in poor countries.[38] Six infectious diseases cause 90 percent of annual worldwide deaths from communicable diseases: acute respiratory infections (3.5 million deaths), AIDS (2.3 million), diarrheal diseases (2.2 million), tuberculosis (1.5 million), malaria (1.1 million), and measles (0.9 million).[38] As a consequence of the AIDS epidemic, some 29 African countries have experienced substantial increases in death rates and substantial declines in average life span. By 2010–2015, life expectancy is projected to decline by 17 years, on average, in the nine hardest-hit countries.[20,39] In Botswana and Zimbabwe, more than 20 percent of the adult population is HIV positive.[20]

Poverty, lack of education, and social and economic factors are powerful, if indirect, correlates of health status. Wealthy nations provide environments that offer protection against infectious diseases through preventive measures such as vaccination, water purification, sanitary sewage disposal, and control of insect vectors. Wealthier nations and individuals can better afford to pay for needed preventive and curative health services. Higher levels of education, especially among women, are also associated with low fertility and good health—the well educated are better equipped to stay healthy and obtain needed health care services.[34,40–43] It is reasonably well established that the families in developing countries with the smallest number of children usually have the highest incomes and the healthiest and best-educated children. Therefore, to the extent that rapid population growth and large family size hamper economic development by perpetuating poverty, high growth rates also contribute to poor health.[25,44–46]

Developing countries that have established strong family planning programs and have successfully slowed rapid population growth have fared much better economically than countries that have neglected the population issue. The Asian economic "tigers"—South Korea, Thailand, Malaysia, and Taiwan—have a 30-year history of supporting family planning and an average of about two children per family. This has benefited the health of their people both by fostering economic development and by establishing a healthy pattern of reproduction.[45–47]

Facing the Challenges of Poor Health, Rapid Population Growth, and High Consumption Levels

In 1994, demographer John Bongaarts[48] disaggregated the sources of future population growth in developing countries into three categories: 49 percent will come from momentum caused by the population's young age structure (the result of previous high fertility); 33 percent will come from unwanted fertility (i.e., births to those who wish to stop childbearing but who are not using contraception); and only 18 percent will come from high desired family size (i.e., desiring more than an average of two children). The fact that most couples in developing countries want small families bodes well for the success of family planning programs in those countries. However, family planning must be accessible. Meeting the family planning needs of the 100–120 million women in developing countries who wish to limit their childbearing but lack access to adequate information and services would lower the TFR from the current 3.2 halfway to the TFR of 2.1 needed for population stabilization.[9,48]

Participants at the 1994 United Nations International Conference on Population and Development made a collective commitment to improve women's status and to make family planning and a limited array of reproductive health services universally available in developing countries by the year 2015.[2] Their emphasis on reproductive health recognized the reality that in developing countries, 25 percent to over 50 percent of treatable or preventable diseases among women aged 15–49 years are related to reproduction. Typically the largest share is associated with pregnancy, unsafe abortion, and childbirth. In some countries, AIDS and other STDs predominate.[44,49–50] Over a woman's lifetime, the risk of dying from pregnancy-related causes is about 1 in 16 in Africa, 1 in 65 in Asia, 1 in 130 in Latin America, but only 1 in 6,000 in the United States and 1 in 10,000 in Northern Europe. A broader array of reproductive health care services, including safe abortion, prenatal care, and the ability to deal with the complications of childbirth, could prevent many deaths.[51–55]

Family planning is necessary to allow a pattern of healthy childbearing. Eliminating childbearing among teenagers, women over 35, and women who have already had four children, and increasing intervals between births to at least two years, would avoid about 25 percent of the 585,000 maternal deaths each year. If women in poor countries bore only two children, the annual number of maternal deaths would be reduced by close to 50 percent.[40,51,53–55]

The safe pattern of childbearing that reduces risk among women also lowers the risk of death among infants and children. Currently each year in developing countries some 11 million children do not survive their first five years of life.[56] Establishing a healthy pattern of childbearing could be expected to reduce infant and child mortality by about 20 percent to 25 percent.[40] For example, in Kenya, a typical developing country, an interval of 18 months or less between births results in a risk of infant death that is twice the risk associated with a longer interval.[40,57–60]

Reproductive health programs that include but also go beyond family planning and safe childbirth services are needed to address domestic violence, which involves up to one in three women,[61] and STDs, which are responsible for 333 million new cases of infection throughout the world each year.[62] Family planning and maternity care programs can serve as a starting point for services that address these problems because they serve the same population of young, sexually active women who are most at risk of exposure to STDs and domestic violence.[60–62]

The cost of family planning and reproductive health services recommended for developing countries at the United Nations International Conference on Population and Development was estimated to be $17 billion annually by the year 2000.[2] It was agreed that two-thirds of this total cost should come from developing countries—an expenditure of less than five cents weekly per person living in these countries—and that one-third should come from donor countries—an expenditure of less than ten cents weekly per person living in developed countries. Unfortunately, developing countries are spending only about $5 billion annually, less than 50 percent of their financial target of $11.3 billion, and donor nations are providing only about $1.4 billion in grants, less than 25 percent of their $5.7 billion goal.[16,17]

If we are able to summon the political will to make good reproductive health care, including family planning and safe abortion, widely available, and if we make reasonable progress in educating women and improving their status, population growth is likely to decline to manageable levels.[51,63,64] In Thailand between 1970 and 1987, a voluntary family planning program, stressing cooperation between public and private sectors, brought about an increase in contraceptive use from about 10 percent to 67 percent of couples.[63] As a result, the average number of children per woman fell from 6.2 to 2.2. Important reasons for the program's success included use of the injectable contraceptive Depo-Provera, the distribution of oral contraceptives by nonphysicians, and strong government support of the program.

Better reproductive health care in poor countries, however, will not be enough to save our natural systems. Both developed and developing countries must introduce economic systems and new technologies that are more efficient, generate less waste, and require less consumption of natural resources.[22,23,30,32,65,66] With the world increasingly seeking economic development through market-based policies, it is imperative that governments and the private sector integrate strategies into economic life that will protect the environment. The way forward to economic progress with more efficiency and less consumption is clear in many sectors, and research can bring additional advances.[30]

The limitation of greenhouse-gas emissions, critical to climactic stabilization, can be addressed by less reliance on and more efficient use of fossil fuels. Further development of wind, geothermal, photovoltaic, and other ecofriendly sources of energy is needed. Carbon emissions can be reduced by preserving forest resources through increased use of recycled paper and wood substitutes. These and other measures to slow the rapid decline in biodiversity are needed. The protection of habitats in ecologically threatened "hot spots" is one promising approach.[30,32,66]

Governments and international development agencies should eliminate environmentally unsound development projects and subsidies for a large array of ecologically unsound practices and products. Policies needing reform include those related to tobacco, mineral production, logging, transportation, agriculture, fisheries, livestock, energy use, waste disposal, control of pesticides and other toxic substances, air quality, and use of land and water resources.[67] Efforts to address the environmental impact of consumption must give attention to the damage and waste caused by conflict and worldwide outlays for military activities, estimated at $700 billion annually.[27]

Of crucial importance is the path of economic development that is traversed by poor countries. China, with a population of 1.2 billion, has experienced an economic expansion of two-thirds since 1990 and a corresponding increase in consumption of many resources.[30] It has surpassed the United States in consumption of grain, meat, fertilizer, steel, and coal. If China's per capita oil consumption equaled that of the United States, the Chinese would consume 80 million barrels a day, far outstripping the daily world production of 60 million barrels. Social and economic progress in China and other developing countries is necessary, but, according to Brown and colleagues, 76 these countries must bypass what the West has done and show how to build environmentally sustainable economies. Unfortunately, many rapidly industrializing countries are proceeding with little regard for the environment.[30]

Reforming our economies and industries will be technically difficult, costly, and time consuming. Measures that will help slow population growth are relatively less expensive and are very effective. Potts has pointed out that "all societies with unconstrained access to fertility regulation experience a rapid decline to replacement levels of fertility, and often lower."[68] Our future well-being depends on increased access to family planning and reproductive health services in developing countries and decreased consumption by people in wealthy countries. We must develop and adopt more efficient technology for industrial production in all countries. Our governments, the private sector, and individuals must work together to devise and adopt new patterns of sustainable economic behavior and to support and enable voluntary and responsible family planning. The challenge is to meet the needs of today's populations without compromising the welfare of future generations.

References

1. United Nations Conference on Environment and Development, 1992 Jun 1–12; Rio de Janeiro. Washington: Council on Environmental Quality, 1992.

2. *Reproductive rights and reproductive health.* International Conference on Population and Development, 1994 Sept 5–13; Cairo. New York: United Nations, 1995.

3. Union of Concerned Scientists. *World scientists' warning to humanity.* Cambridge, MA: Union of Concerned Scientists, 1992 Nov.

4. *Population summit of the world's scientific academies.* Washington, DC: National Academy Press, 1994.

5. Mazur LA, editor. *Beyond the numbers: A reader on population, consumption and the environment.* Washington, DC: Island Press, 1994.

6. Cohen JE. *How many people can the earth support?* New York: WW Norton, 1995.

7. *2001 world population data sheet.* Washington, DC: Population Reference Bureau; 2001. Available: http://www.prb.org/pubs/wpds2001

8. Freedman R, Blanc AK. Fertility transition: An update. In: Blanc AK, editor. *Demographic and health surveys world conference.* vol. 1. Columbia, MD: IRD/Macro International, 1991.

9. Bongaarts J, Mauldin WP, Phillips JF. The demographic impact of family planning programs. *Stud Fam Plann* 1990; 21:299–310.

10. *1987 world population data sheet.* Washington, DC: Population Reference Bureau, 1987.

11. *World population prospects: The 2000 revision, highlights.* New York: United Nations, 2001.

12. *World population projections to 2150.* New York: United Nations, 1998.

13. Bongaarts J. Demographic consequences of declining fertility. *Science* 1998; 282:419–420.

14. McDevitt TM. *World population profile: 1998.* Washington, DC: US Census Bureau, 1999. Available: http://www.census.gov/ipc/www/wp98.html

15. Conly SR, Chaya N, Helsing K. *Contraceptive choice: Worldwide access to family planning.* Washington, DC: Population Action International, 1997. Available: http://www.populationaction.org/programs/rc97.htm

16. Conly SR, Speidel JJ. *Global population assistance: A report card on the major donor countries.* Washington, DC: Population Action International, 1993.

17. Conly SR, De Silva S. *Paying their fair share? Donor countries and international population assistance.* Washington, DC: Population Action International, 1999. Available: http://www.populationaction.org/programs/dac98/dac download.htm

18. *Population projections of the United States by age, sex, race, and Hispanic origin: 1995 to 2050.* Washington, DC: US Census Bureau, 1996. Current Population Reports P25–1130. Available: http://www.census.gov/prod/1/pop/p25–1130

19. *Annual demographic survey.* Washington, DC: US Census Bureau, 1997 Mar. [suppl to Current Population Survey].

20. *The world at six billion.* New York: United Nations (Population Division), 1999. Available: http://www.un.org/esa/population/sixbillion.htm

21. *Components of population growth: Canada, the provinces and territories, July 1, 1998–June 30, 1999.* Ottawa: Statistics Canada, 1999. Available: http://www.statcan.ca/english/pgdb/people/population/demo33a.htm

22. Ehrlich PR, Ehrlich AH. *The population explosion.* New York: Simon and Schuster, 1990.

23. Green C. The environment and population growth: Decade for action. *Population Reports,* Series M, no. 10. Baltimore: Johns Hopkins School of Hygiene and Public Health, 1992.

24. World Resources Institute. *World resources 1998–99. A guide to the global environment: Environmental change and human health.* New York: Oxford University Press, 1998.

25. Ahlburg DA. Population growth and poverty. In: Ahlburg DA, Kelley AC, Mason KD, editors. The impact of population growth on well-being in developing countries. New York: Springer–Verlag, 1996.

26. *Why population matters.* Washington, DC: Population Action International, 1998. Available: http://www.populationaction.org/why_pop/whypop.htm

27. Brown LR, Renner M, Flavin C. *Vital signs, 1998.* Washington, DC: Worldwatch Institute, 1998.

28. World Resources Institute. *World resources 1994–95: People and the environment.* New York: Oxford University Press, 1994.

29. Bender W, Smith M. Population, food and nutrition. *Popul Bull* 1997; 51:4.

30. Brown LR, Flavin C, French H. *State of the world 1998.* Washington, DC: Worldwatch Institute, 1998.

31. Our demographically divided world: Rising mortality joins falling fertility to slow population growth. [Press release.] Washington, DC: Worldwatch Institute, 1999 Apr 10. Available: http://www.worldwatch.org/alerts/990408.html

32. Brown, LR, Flavin C, and French H. 2000. *State of the world 2000.* Washington: Worldwatch Institute.

33. Brown LR, Gardner G, Halweil B. *Beyond Malthus: Nineteen dimensions of the population challenge.* Worldwatch Environmental Alert series. New York: WW Norton, 1999.

34. World Bank. *World development report 1993: Investing in health.* New York: Oxford University Press, 1993.

35. Brown LR, Abramovitz J, Bright C, Flavin C, Gardner G, Kane H, et al. *State of the world 1996.* Washington, DC: Worldwatch Institute, 1996.

36. *Climate change, state of knowledge.* Washington, DC: Office of Science and Technology, 1997.

37. Paltz JM, Epstein PR, Burke TA, Balbus JM. Global climate change and emerging infectious diseases. *JAMA* 1996; 275:217–223.

38. *Removing obstacles to healthy development: World Health Organization report on infectious disease.* Geneva: World Health Organization, 1999. Available: http://www.who.int/infectious-disease-report/index-rpt99.html

39. *The UN AIDS Report.* Geneva: UNAIDS, 1999. Available: http://www.unaids.org/publications/documents/unaids/index.html

40. Healthier mothers and children through family planning. *Popul Rep J* 1984; May-Jun(27):J657–J696

41. *The health rationale for family planning: Timing of births and child survival.* New York: United Nations, Population Division, Dept. for Economic and Social Information and Policy Analysis, 1994.

42. *Human development report 1996.* New York: United Nations Development Programme, 1996. Overview available: http://www.undp.org/hdro/96.htm

43. *The world health report 1999.* Geneva: World Health Organization, 1999. Available: http://www.who.int/whr/1999/en/report.htm

44. Pappas G, Queen S, Hadden W, Fisher G. The increasing disparity in mortality between socioeconomic groups in the United States, 1960 and 1986. *N Engl J Med* 1993; 329:103–109.

45. Population Crisis Committee. Population growth and economic development: The new policy debate. *Population* 1985 Feb; 14:1–8.

46. Knodel J, Havanon N, Sittitrai W. *Family size and the education of children in the context of rapid fertility decline.* Ann Arbor: Population Studies Center, University of Michigan, 1989. Rep no 89–155.

47. Bloom DE, Williamson JG. Demographic transitions and economic miracles in emerging Asia. *World Bank Econ Rev* 1998; 12:419–455.

48. Bongaarts J. Population policy options in the developing world. *Science* 1994; 263: 771–776.

49. Duke RC, Speidel JJ. The 1991 Albert Lasker Public Service Award. Women's reproductive health: A chronic crisis. *JAMA* 1991; 266:1846–1847.

50. Speidel JJ. Population: What does it mean to health? *Physicians for Social Responsibility Quarterly* 1993; 3(4):155–165.

51. Starrs A. *Preventing the tragedy of maternal deaths: A report on the International Safe Motherhood Conference, Nairobi, Kenya, February 1987*. Washington, DC: World Bank, 1987.

52. *Maternal mortality rates: A tabulation of available information*. Geneva: World Health Organization, 1986.

53. *Revised 1990 estimates of maternal mortality: A new approach by WHO and UNICEF*. Geneva: World Health Organization, 1996.

54. Winikoff B, Sullivan M. Assessing the role of family planning in reducing maternal mortality. *Stud Fam Plann* 1987; 18:128–143.

55. Alan Guttmacher Institute. *Sharing responsibility: Women, society and abortion worldwide*. New York: Alan Guttmacher Institute, 1999.

56. Grant JP. *The state of the world's children*. New York: Oxford University Press for UNICEF, 1991.

57. Conly SR. *Family planning and child survival. The role of reproductive factors in infant and child mortality: An analysis*. Washington, DC: Population Crisis Committee, 1991.

58. Acsadi GT, Johnson-Acsadi G. *Optimum conditions for childbearing*. London: International Planned Parenthood Federation, 1986.

59. Shane B. *Family planning saves lives*. 3rd ed. Washington, DC: Population Reference Bureau, 1997.

60. Khanna J, Van Look PFA, Griffin PD, editors. *Reproductive health: A key to a brighter future: Biennial report 1990–1991*. Geneva: World Health Organization, 1992.

61. Heise L, Ellsberg M, Gottemoeller M. Ending violence against women. *Popul Rep* Series L 1999 Dec; 1–43

62. Murray CJL, Lopez AD, editors. *Health dimensions of sex and reproduction: The global burden of sexually transmitted diseases, HIV, maternal conditions, perinatal disorders, and congenital anomalies*. Vol. 3 of *Global burden of disease and injury*. Boston: Harvard School of Public Health on behalf of the World Health Organization and the World Bank, 1998.

63. Bennett A, Frisen C, Kamnuansilpa P, McWilliam J. How Thailand's Family Planning Program reached replacement level fertility: Lessons learned. Washington: US Agency for International Development, 1990. [Population technical assistance project, occasional paper no. 4.]

64. *High stakes: The United States, global population and our common future. A report to the American people from the Rockefeller Foundation*. New York: Rockefeller Foundation, 1997.

65. Cohen JE. *How many people can the earth support?* New York: WW Norton, 1995.

66. Brown LR, Flavin C, French H. *State of the world 1999*. Washington, DC: Worldwatch Institute, 1999.

67. Brown LR, Flavin C, French H. *State of the world 1997*. Washington, DC: Worldwatch Institute, 1997.

68. Potts M. Sex and the birth rate: human biology, demographic change, and access to fertility-regulation methods. *Population and Development Review* 1997; 23(1):8.

Global Climate Change and Health

Andrew Haines, Anthony J. McMichael, and Paul R. Epstein

6

Greenhouse gases, naturally present at low concentrations in the lower atmosphere, keep the earth's mean surface temperature at around 15°C. Without this trapping of heat ("radiative forcing"), the mean air temperature would be −18°C and the earth would freeze. Figure 6.1 illustrates the mechanism of the greenhouse effect. The atmospheric concentrations of greenhouse gases have been increasing since the early industrial revolution, owing principally to humankind's rapidly increasing combustion of fossil fuels, along with increases in deforestation, irrigated agriculture, animal husbandry, and cement manufacture.[1]

The concentrations of the main greenhouse gases (carbon dioxide, methane, nitrous oxide) reached their highest values over the last millennium during the 1990s (figure 6.2).[2] The warming effect of these gases is much larger than the atmospheric cooling caused by aerosols, such as short-lived sulfates from sulfur dioxide, and the depletion of stratospheric ozone. There was little change in solar energy output during the second half of the twentieth century, but there is strong evidence that on a global scale the earth's climate is now warmer than at any time during the 140 years since records began, due to the accumulation of greenhouse gases in the atmosphere (figure 6.3).[3] Even more dramatically, the 1990s appear to be the warmest decade in the last 1,000 years.

A number of other changes in the earth's biophysical systems in the recent past are consistent with climate change. For example, global mean sea level has risen by between 10 and 20 cm from 1860 to 2000, and snow

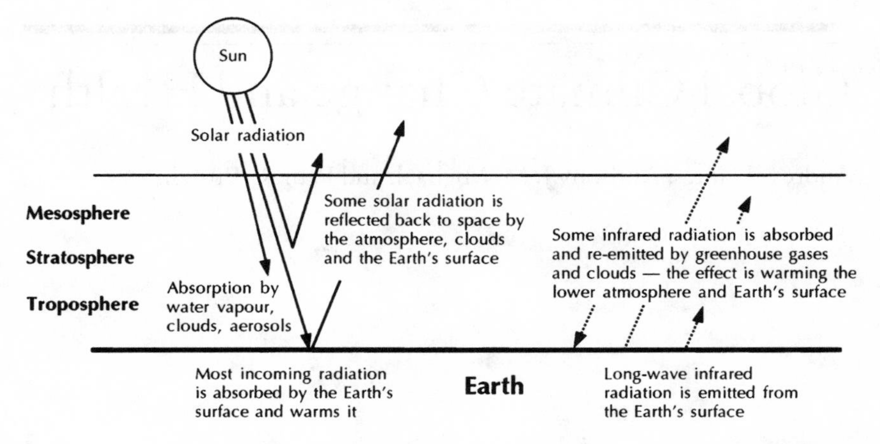

Figure 6.1 Mechanisms of the greenhouse effect. Adapted with permission from WHO.

cover has decreased by 10 percent since the 1960s, with evidence of thawing of permafrost. The third assessment report of the United Nations Intergovernmental Panel on Climate Change (IPCC) comprehensively reviewed the science of climate change and its potential impacts.[2,4] The panel concluded that, using a range of plausible emissions scenarios, global average surface temperature is projected to rise by 1.4°C to 5.8°C between 1990 and 2100. This is about a 2 to 15 times greater increase than any observed during the last 100 years and will probably be unequaled during at least the last 10,000 years. Land areas are likely to warm faster than average. This assessment is derived from projections made by computer-based global climate models that combine, through simultaneous equations within a three-dimensional global grid, the atmospheric and oceanic processes that occur in response to increased greenhouse gases and the resulting rise in radiative forcing in the lower atmosphere.

Although the current generation of global climate models cannot forecast the precise spatial and temporal pattern of changes in climate means and variability with global warming, extreme weather events such as drought, floods, and storms are likely to become more frequent and more intense in the future. Indeed, with warming ocean surfaces and the fact that each increase of 1°C in temperature enables the atmosphere to hold 6 percent more water vapor, the resulting intensification of the hydrological cycle corresponds to evidence in the United States and other

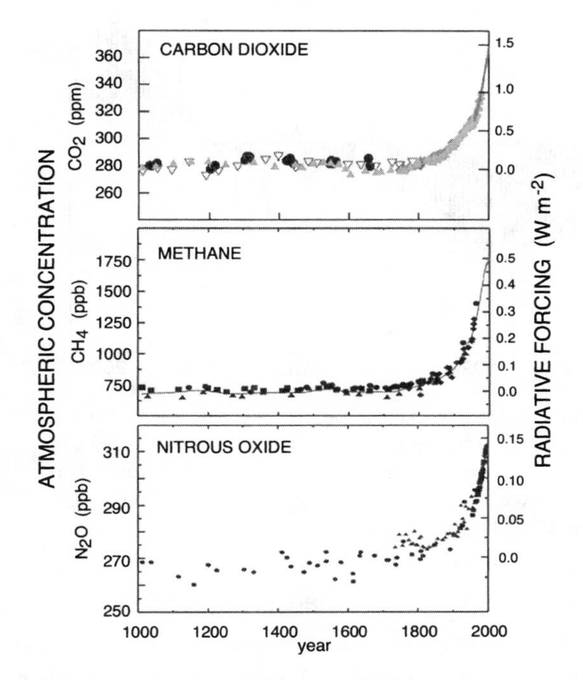

Figure 6.2 Records of past changes in atmospheric composition demonstrate the rapid rise in greenhouse gases that is attributable primarily to industrial growth since 1750. The ice-core and firn data for several sites in Antarctica and Greenland (shown by different symbols) are supplemented with the data from direct atmospheric samples over the past few decades (shown by the line for CO_2 and incorporated in the curve representing the global average of CH_4). The estimated positive radiative forcing of the climate system from these gases is indicated on the right-hand scale. Adapted from Ref. 2.

nations of an increase in heavy rain events and prolonged droughts in the twentieth century.[5] There is evidence that, since the mid–1970s, El Niño events have increased in frequency, magnitude, and duration[6] and that climate change may alter the frequency and magnitude of the El Niño Southern Oscillation (ENSO) cycle.[7]

Recent evidence demonstrates anthropogenically driven warming of the deep oceans down to 3 km.[8,9] Additionally, extensive warming at high latitudes is apparently melting polar and Greenland ice.[10] North Atlantic climate change occurring since 1950 appears linked to progressive

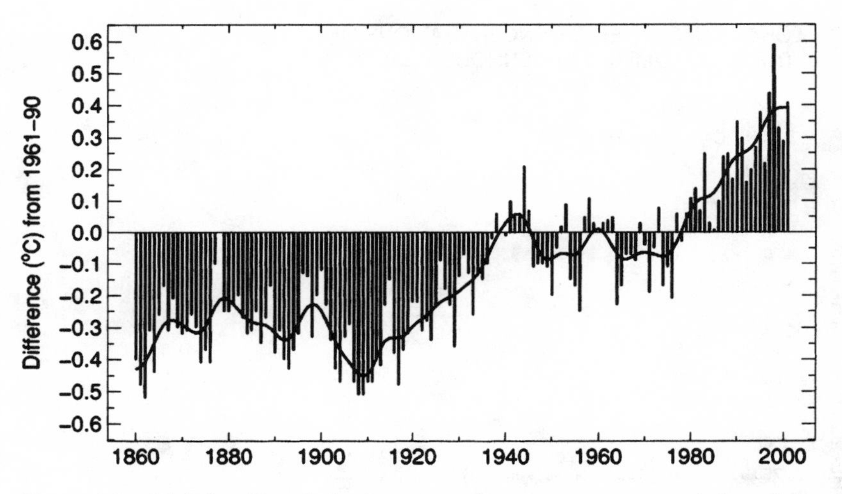

Figure 6.3 Global and hemispheric near-surface temperatures, consisting of annual differences from 1961–1990 normals. This information is based on regular measurements of air temperature at land stations and on sea surface temperatures measured from ships and buoys. Reprinted with permission from the Hadley Centre for Climate Prediction and Research, UK Meteorological Office.

warming of tropical sea surface temperatures influencing the North Atlantic oscillation that drives weather patterns in Europe, Eurasia, and the Northeastern United States. Greater departure from norms may also indicate systemic instability of the climate regime,[11] increasing the potential for abrupt climate change. Most climate models project a weakening of the thermohaline circulation (the global circulation system in the world oceans). Since, for example, the poleward flow of warm water increases the temperature of Northwestern Europe by up to 10°C, any weakening or cessation of the circulation could lead to pronounced regional cooling.[2]

Signals of Climate Change

Global warming is projected to increase both ambient temperatures and rainfall at high latitudes and high elevations. The growing season of crops, particularly at higher latitudes, has lengthened. For example, across Europe, the increase has been around 11 days between 1959 and 1993. There is evidence that breeding, flowering, and migration have been brought forward. Between 1936 and 1998, for instance, 36 flowering species in the

Central United States advanced by an average of 7.3 days.[4] The migration of plants to higher altitudes has been documented on numerous peaks in the European Alps, Alaska, the Sierra Nevada (in the United States), and New Zealand.[12] These botanical trends, indicative of warming, have accompanied other physical changes, such as the retreat of montain glaciers in Argentina, Peru, Alaska, Iceland, Norway, the Swiss Alps, Kenya, the Himalayas, Indonesia, and New Zealand.[13] Since 1970, the lowest level at which freezing occurs has ascended about 150 m higher in mountains in tropical latitudes (from 30°N to 30°S latitude), which is equivalent to 1° C warming.[14]

Meanwhile, there have been reports that both insects and insect-borne diseases (including malaria and dengue fever) have been increasing at high altitudes in Africa, Asia, and Latin America.[15–19] A number of factors may be implicated, including deforestation, population growth, and movements and breakdown in public health systems. Although it is not yet possible to attribute these increases conclusively to climatic change,[20–23] a climatic influence is plausible, and the emerging pattern is compatible with the botanical and physical evidence of warming at high altitudes.[24] There is some evidence that the northern limit of distribution of the tick *vector (Ixodes ricinus)* and tick density increased in Sweden between the early 1980s and the mid–1990s.[25] In Europe, ticks transmit Lyme disease and tick-borne encephalitis.

The cost of global weather-related catastrophic losses has risen tenfold from the 1950s to the 1990s, corrected for inflation. Including other (noncatastrophic) weather events doubles the cost. Increasing losses cannot be fully explained by other factors, such as demographic change or increases in insured value, and the number of weather-related catastrophic events has risen approximately three times faster than those of non-weather-related events.

The challenge for epidemiologists is to pick the settings, populations, and health outcomes where we have the best chance of detecting early effects and attributing these effects to climate change. Research opportunities must be sought where the necessary information exists. Since climate change is, essentially, an incremental phenomenon that already extends over several past decades, time-series data on outcomes and confounders over a similar period are required. Early impacts are likely to be seen most clearly where the exposure-outcome gradient is steep and adaptive capacity is weak; they will be most easily attributable to climate change when there are few competing explanations.

The best bets for epidemiological studies of early effects include the following three possibilities:

1. Vector-borne diseases may be relatively sensitive indicators, since transmission involves intermediate organisms, such as mosquitoes, open to environmental influences.
2. Intestinal infections (food poisoning) show strong seasonal patterns (suggesting a powerful effect of climate variability) and have been routinely reported for many years (although the data are known to be incomplete).
3. Deaths, injuries, and illnesses caused by extreme events (such as heat waves, cold spells, floods, and storms) satisfy the condition of "few competing explanations," but, in many populations, it may be difficult to distinguish the climate-change signal from much stronger mitigating effects of social and economic development.

For epidemiologists, this presents an emerging opportunity for empirical research. A major task over the next five to ten years will be to initiate well-directed local epidemiologic studies and well-coordinated regional and international data-collection systems to achieve early detection of presumptive now-occurring climate-influenced changes in population health outcomes.

The Range of Potential Health Impacts

A change in world climate would have wide-ranging, mostly adverse, consequences for human health.[26–28] Most anticipated health impacts would entail increased rates of illnesses and death from familiar causes (table 6.1). However, the assessment of future health outcomes refers to climatic-environmental conditions not previously encountered. Such conditions, particularly in conjunction with other global environmental changes now occurring (e.g., deforestation) may also increase the likelihood of unfamiliar health outcomes, including the emergence of "new" infectious-disease agents.[29] The 1997–1998 El Niño event brought surprises: Indonesia and Brazil experienced widespread respiratory illness due to haze from uncontrolled burning of tropical forests.[30] Following Hurricane Mitch in Central America in November 1998, deforested areas experienced increased flooding and landslides, the aftermath spawning "clusters" of water-, insect-, and rodent-borne diseases (cholera, malaria, dengue fever, and leptospirosis).

The potential health impacts of global warming can be broadly classified as direct or indirect (see table 6.1) . The former category refers to the direct impact of extremes in local weather conditions. Epidemiologic studies and public health data have identified how thermal stresses (in-

Table 6.1 Mediating processes and direct and indirect potential effects on health of changes in temperature and weather

Mediating process	Health outcome
Direct effects	
Exposure to thermal extremes	Changed rates of illness and death related to heat and cold
Changed frequency or intensity of other extreme weatherevents	Deaths, injuries, psychological disorders; damage to public health infrastructure
Indirect effects	
Disturbances of ecological systems:	
Effect on range and activity of vectors and infective parasites	Changes in geographic ranges and incidence of vector-borne disease
Changed local ecology of water-borne and food-borne infective agents	Changed incidence of diarrheal and other infectious diseases
Changed food productivity (especially crops) through changes in climate and associated pests and diseases	Malnutrition and hunger, and consequent impairment of child growth and development
Sea-level rise with population displacement and damage to infrastructure	Increased risk of infectious disease, psychological disorders
Biological impact of air pollution changes (including pollens and spores)	Asthma and allergies; other acute and chronic respiratory disorders and deaths
Social, economic, and demographic dislocation through effects on economy, infrastructure, and resource supply	Wide range of public health consequences: mental health and nutritional impairment, infectious diseases, civil strife

Source: Reprinted from McMichael and Haines (reference 27) with permission from BMJ Publishing Group.

cluding heat waves) and weather disasters can result in serious illness, injuries, and death. Estimating the consequences of indirect effects poses more of a challenge because those impacts typically result from changes in complex processes. They include alterations in the transmission of vector-borne infectious diseases, alterations in water quality and quantity, and changes in the productivity of agroecosystems,[31] with the potential for displacement of vulnerable populations resulting from local declines in food supply or sea-level rise. The range of likely health impacts can be assessed, in part, by studying the consequences of local climatic variability, including short-term trends.

One useful, although limited, analogue of future climatic change is the ENSO cycle, which affects temperature, precipitation, and extreme events (e.g., storms) in many parts of the world. The ENSO cycle influences, often strongly, the incidence of various infectious diseases in many parts of the world: malaria in northeastern Pakistan, Sri Lanka, Colombia, and Venezuela; Murray Valley encephalitis and epidemic polyarthritis (Ross River virus) in Australia; and dengue fever in the South Pacific.[32–37] Historical analyses showed that the risk of a malaria epidemic increased fivefold in the semiarid Punjab during the year following an El Niño and fourfold in southwestern Sri Lanka during the El Niño year. El Niño events are also strongly associated with the numbers of people affected globally by natural disasters, particularly droughts, that cause major harm to human health.[38] A recent review has documented a range of health impacts that may be affected by the ENSO cycle.[39]

Direct Health Impacts

Thermal extremes provide a clearly relevant form of climatic variation for study, because global warming is projected to increase the frequency of heat waves and decrease the frequency of winter cold spells. Modeling based on previous studies of mortality associated with heat waves in specified urban populations in the United States has indicated that the rate of deaths related to heat waves might increase substantially by the year 2050, particularly if little acclimatization to warmer weather occurs.[40] A study of ten Canadian cities suggested that in Montreal, for example, heat-related deaths would increase from 70 per annum to 240–1,140 in an "average" summer in 2050 without acclimatization.[41] The investigators suggested that some acclimatization of populations might occur in Montreal and

Toronto but probably not in Ottawa. Because the conversion of nitrogen oxide to ground-level ozone (smog) is temperature dependent, a projected fivefold increase in the number of hot days with temperatures above 30°C could lead to increases in the number of days with concentrations of ground-level ozone considered a risk to health for sensitive individuals.[42]

The association between mortality and temperature is J-shaped in many countries. The relation between increased mortality and low temperatures is more complex than that between increased mortality and high temperatures; thus, the degree to which cold-related deaths in temperate countries may decline with global warming is unresolved. Excess winter mortality is higher in some countries with temperate climates (e.g., the United Kingdom) than in those with very cold winters, probably because of insufficient fuel and inadequate housing and winter clothing.[43,44] Although much of the winter excess in mortality is due to increases in cardiovascular events, some may be due to viral infections arising from increased crowding during winter.[45] If there are substantial changes in the ocean circulation, parts of the world (e.g., Northwestern Europe) experiencing reductions in warm ocean currents could suffer increases in cold-related deaths, despite overall increases in global temperature.

Indirect Health Impacts

In addition, disproportionate warming at high latitudes, at high altitudes, during winter, and at night could produce disproportionate impacts. Canada, for example, could experience a greater relative increase in heat waves and conditions conducive to outbreaks of vector-borne infectious diseases than nations at lower latitudes. In Canada, malaria disappeared at the end of the nineteenth century, although in 1998 cases of locally transmitted malaria were reported in Toronto.[29] But though increased temperatures may result in conditions suitable for the reintroduction of malaria, the existence of effective public health programs will be the main determinant of the existence and extent of such infections. Many of the encephalitides in North America, including St. Louis and La Crosse encephalitis and western, eastern, and Venezuelan equine encephalomyelitis, are transmitted by mosquitoes. Although the mosquito life span tends to diminish if temperatures rise excessively, viral maturation rates increase with temperature, within the viable ranges of the mosquitoes and pathogens. It has been suggested that, as a result of climate change, there could be a northward shift in western equine and St. Louis encephalitis (SLE), with the disappearance of the former in southern endemic regions.[46] Urban outbreaks of SLE and West Nile virus infection may be associated with prolonged drought.[47]

Complex, integrated mathematical models are used to estimate the likely effects of climate change on vector-borne diseases. These highly aggregated models are in the early stages of development and do not take into account local environmental and ecological circumstances.[48] Nevertheless, they are useful for forecasting the broad direction and potential magnitude of future change. The majority of such models project substantial increases in the zones where malaria and dengue fever could be transmitted. For example, one estimate using a biological model suggested that the proportion of the world population living within the potential malaria transmission zone would increase from around 45 percent in the 1990s to around 60 percent by 2050. But there may be a decrease in the transmissibility of schistosomiasis because of global warming of water and some regional drying.[49–51] Another study using a statistical-empirical approach assessed the likely changes in people living in the actual, as distinct from the potential, transmission zone of malaria.[52] It concluded that there would be less net change by 2050. For instance, under the "medium–high" scenario, there might be an increase in exposure of 23 million people, but under the "high" scenario, a decrease in exposure of 25 million people. However, for at least two reasons, this study may have underestimated the effects of climate change on malaria. First, the current geographic distribution of malaria, which was used as the basis for the estimations, does not reflect the full range of locations where climate permits malaria transmission. Second, average temperatures were used, whereas malaria transmission may be more sensitive to changes in minimum temperature. The actual changes in the incidence of malaria, dengue fever, and other vector-borne diseases would, of course, depend on many factors, including future patterns of social development, land use, and urban growth, and the effectiveness of preventive measures, such as vector control and vaccination. A projected decrease in the incidence of a specified vector-borne disease as a result of climate change does not necessarily imply a beneficial effect on human health, since changes in temperature and humidity that affect vector populations may also have other adverse effects on health.

The growth of algae in surface waters, estuaries, and coastal waters is sensitive to temperature.[53,54] About 40 of the 5,000 species of marine phytoplankton (algae) can produce biotoxins, which may reach human consumers through shellfish. Warmer sea temperatures can encourage a shift in species composition of algae toward the more toxic dinoflagellates.[55] Upsurges of toxic phytoplankton blooms in Asia are strongly correlated with the ENSO cycle.[56]

Algal blooms may potentiate the transmission of cholera. Electron microscopy has shown that algae and the zooplankton that feed on them provide a natural refuge for *Vibrio cholerae,* where, under normal conditions, the bacteria exist in a nonculturable, dormant state. An increase in sea surface temperature, along with high nutrient levels (eutrophication) that stimulate algal growth and deplete oxygen, can activate the blooms and vibrios. Sea surface temperature in the Bay of Bengal is correlated with algal blooms and outbreaks of cholera in Bangladesh.[57] Climate variability and change may thus influence the introduction of cholera into coastal populations. *V. cholerae* occur in the Gulf of Mexico and along the East Coast of North America.

Heavy rainfall may lead to outbreaks of cryptosporidiosis,[58] which causes severe diarrhea in children and can cause death in immunocompromised individuals. Rodent populations are also influenced by climate anomalies. Prolonged droughts deplete rodent predators (owls, snakes, and coyotes), whereas rains provide new food supplies. These dynamics apparently contributed to the 1993 outbreak of hantavirus pulmonary syndrome in the Southwestern United States[59,60] (see chapter 7) and may have contributed to recent outbreaks of that disease in Argentina, Bolivia, Chile, Canada, and Paraguay.[61–64] In North America and Europe, large deer populations (with few predators) and warm winters, allowing overwintering of tick populations at higher latitudes,[37] could increase the range of Lyme disease.

Climate change could also affect food production. Modeling studies suggest that yields of cereal grain, the main food commodity on a global scale, would increase at high latitudes. Declines would be concentrated in low-latitude regions, including Africa, the Middle East, and India, where food insecurity often already exists.[31,65] There is a range of estimates of the risk of hunger, reflecting different assumptions about future population growth, international trade, and adaptive agricultural technology. Such estimates, however, do not include the likely additional influence of extreme weather events or of increases in agricultural pests and pathogens, which could be substantial.[66]

Although studies disagree about the impact of limited climate change (up to 2.5°C) on food prices, there appears to be growing evidence that a mean global temperature rise of more than 2.5°C would result in an increase in food prices, with obvious implications for access to food supplies by the poor.[4]

It has been suggested that the projected 50 percent increase in the global population and demand for diets richer in meat will double global food demand by 2050. This will create challenges. These challenges will rival and interact with climate change, leading to increases in nitrogen- and phosphorus-driven eutrophication of terrestrial, freshwater, and near-shore marine ecosystems. Together with habitat destruction, these developments in turn will result in simplification of ecosystems, loss of ecosystem services, and accelerated species extinctions.[67]

Accelerated sea-level rise would have a variety of health impacts. The third assessment report of the IPCC gave a range of sea-level rise of 0.09—0.88 m between 1990 and 2100 for the full range of scenarios that they considered. Much larger rises are possible over time. For example, a sustained localized warming of 3°C or 5.5°C over Greenland could result in a sea-level rise of 1 m or 3 m, respectively, over the next 1,000 years because of melting of the ice sheet.[4] For a 40 cm rise in sea level compared with no further rise from the present, the number of people flooded annually would increase from 13 million to around 90 million: 60 percent in South Asia and 20 percent in Southeast Asia. The degree of coastal protection could greatly influence the probability of flooding. If the level of protection is not improved proportionately to GDP growth, the numbers of people flooded annually could more than double. Populations on low-lying islands such as the Maldives, the Marshall Islands, Kiribati, and Tonga and in the deltaic regions of parts of Africa and the United States would also be vulnerable to accelerated rises in sea level and associated increases in storm surges. Salination of coastal farmlands and of freshwater aquifers would cause economic disruption, population displacement, and additional adverse health consequences. In Canada, for example, much of the coast of Prince Edward Island is highly erodable, and shorefront buildings may be threatened along the Gulf of St. Lawrence. Rises in sea level and increased storm surges along the tundra coast of Alaska and Canada are likely to cause erosion and flooding.[65] The Arctic and the Antarctic are, in general, likely to be particularly vulnerable to climate change, resulting, for example, in substantial loss of sea ice and changes in species composition, with implications for indigenous human communities following traditional lifestyles. In addition, loss of ice cover will alter the earth's albedo (reflectivity), thus increasing heat absorption and contributing to climate change.

Vulnerability

Population vulnerability to climate variability and climate change depends on the sensitivity of a given health outcome to climate change and the ability of the population to adapt to new climatic conditions. This capacity is likely to be determined by factors at the level of the individual, the community, and the geographic location. Individual factors may include socioeconomic deprivation (which may also be an important factor at the level of the community), disease status (i.e., those with preexisting disease are more likely to be vulnerable), and demographic factors, such as age. Community factors may include effectiveness of water and sanitation systems, local food supplies and distribution systems, and vector and other disease control programs. Effective early-warning systems could also reduce vulnerability at the community level. Geographic factors may include situation in low-lying coastal areas or in locations where freshwater availability is limited. Environmentally degraded and deforested areas are likely to be more vulnerable to extreme weather events.

Prescriptions

Carbon dioxide has a long residence time in the atmosphere, as do some other greenhouse gases. In addition, there is substantial inertia in the atmosphere/ocean/climate system. Thus, the emissions of our generation will affect future generations for many years to come. Merely stabilizing emissions at current levels will result in a continually rising atmospheric carbon dioxide concentration and temperature. Substantial reductions in emissions will be required to stabilize the atmospheric concentrations of carbon dioxide.

Industrialized nations produce most of the world's greenhouse-gas emissions. Even if these nations achieve the limited reductions agreed to at the 1997 Kyoto Climate Change Convention, global carbon dioxide emissions are likely to increase substantially, with increasing contributions from countries such as China, India, and other developing nations.[68] (At the time of this writing, even the attainment of these limited reductions looks problematic, due to the opposition of the Bush administration in the United States in particular.) To protect their own development prospects, developing countries will therefore need substantial incentives to cut emissions, including the transfer of nonpolluting renewable energy and energy-efficient technologies. Reducing fossil-fuel combustion will also have substantial direct near-term health benefits, such as preventing many thousands of air-pollution-induced deaths annually worldwide from both indoor and outdoor sources, particularly in less developed countries.[69,70] Reducing population growth in the South will facilitate attempts to curb the demand for fossil-fuel energy. Population growth is linked to poverty, and poor populations will experience greater vulnerability to the adverse effects of climate change. Thus, development policies must encompass effective poverty-reduction strategies and access to culturally acceptable methods of family planning and to greenhouse-gas mitigation technologies.

Some degree of global warming now seems certain. Thus, adaptations to climate change will be required, such as housing designs that enhance summertime cooling, the "greening" of inner cities, the strengthening of coastal defenses, and improved control of vector-borne and waterborne diseases. Health-indicator monitoring and disease surveillance should be integrated with the three nascent global observing systems for world climate, oceans, and terrestrial systems.[71] Multidisciplinary research into the identification, understanding, and modeling of health impacts needs support, as do intergovernmental and interagency collaborations to develop health early-warning systems that can facilitate timely, environmentally friendly public health interventions.

Recognizing the wide-ranging potential consequences of climate change for our health and well-being can greatly strengthen the international rationale for reducing greenhouse-gas emissions. Although there is much that is unavoidably complex and uncertain about these large-scale risks to human population health, the need for health professionals to take a health-protecting, precautionary approach that will have multiple health benefits remains clear.[72]

References

1. Houghton JT, Meira-Filho LG, Callander BA, Harris N, Kattenberg A, Maskell K, editors. *Climate change, 1995—the science of climate change: Contribution of Working Group I to the second assessment report of the Intergovernmental Panel on Climate Change.* New York: Cambridge University Press, 1996.

2. Albritton DI, Allen MR, Baede APM, et al., eds. Intergovernmental Panel on Climate Change Working Group I. Summary for Policy Makers, Third Assessment Report. *Climate change 2001: The Scientific Basis.* Geneva, 2001.

3. *Climate change and its impacts.* London: Hadley Centre for Climate Prediction and Research, UK Meteorological Office; 2001. Available: www.metoffice.gov.uk/sec5/CR div/CoP5/contents.html

4. McCarthy JJ, Canziani, OF, Leary NA, Dokken DJ, White KS, eds. Intergovernmental Panel on Climate Change Working Group II. Third Assessment Report. *Climate Change 2001: Impacts, Adaptation, and Vulnerability.* Geneva: 2001.

5. Karl TR, Knight RW, Easterling DR, Quayle RG. Trends in US climate during the twentieth century. *Consequences* 1995; 1:3–12.

6. Trenberth KE, Hoar TJ. The 1990–1995 El Niño–southern oscillation event: Longest on record. *Geophys Res Lett* 1996; 23:57–60.

7. Timmermann A, Oberhuber J, Bacher A, Esch M, Latif M, Roeckner E. Increased El Niño frequency in a climate model forced by future greenhouse warming. *Nature* 1999; 398:694–697.

8. Levitus S, Antonov JI, Wang J, Delworth TL, Dixon KW, Broccoli AJ. Anthropogenic warming of earth's climate system. *Science* 2001; 292:267–270.

9. Barnett TP, Pierce DW, Schnur R. Detection of anthropogenic climate change in the world's oceans. *Science* 2001; 292:270–274.

10. Hoerling MP, Hurrell JW, Xu T. Tropical origins for recent North Atlantic climate change. *Science* 2001; 292:90–92.

11. Taylor K. Rapid climate change. *Am Sci* 1999; 87:320.

12. Pauli H, Gottfried M, Grabherr G. Effects of climate change on mountain ecosystems—Upward shifting of alpine plants. *World Resource Rev* 1996; 8:382–390.

13. Thompson LG, Mosley-Thompson E, Davis M, et al. "Recent warming": Ice core evidence from tropical ice cores with emphasis on Central Asia. *Global and Planetary Change* 1993; 7:145–156.

14. Diaz HF, Graham NE. Recent changes in tropical freezing heights and the role of sea surface temperature. *Nature* 1996; 383:152–155.

15. Loevinsohn M. Climatic warming and increased malaria incidence in Rwanda. *Lancet* 1994; 343: 714–718.

16. Bouma MJ. *Epidemiology and control of malaria in northern Pakistan.* Den Haag: CIP-Gegevens Koninklijke Bibliotheek, 1995.

17. Koopman JS, Prevots DR, Vaca Marin MA, Gomez Dantes H, Zarate Aquino ML, Longini IM. Determinants and predictors of dengue infection in Mexico. *Am J Epidemiol* 1991; 133:1168–1178.

18. Subianto B. *Malaria—Indonesia (Irian Jaya).* Boston: ProMED-mail (Program for Monitoring Emerging Diseases), International Society for Infectious Diseases, 1997 Dec. 23.

19. Tulu A. *Determinants of malaria transmission in the highlands of Ethiopia: The impact of global warming on morbidity and mortality ascribed to malaria* [dissertation]. London: School of Hygiene and Tropical Medicine, 1996.

20. Lindsay SW, Martens WJM. Malaria in the African highlands: Past, present and future. *Bull World Health Organ* 1998; 76:33–45.

21. Malakooti MA, Biomndo K, Shanks DG. Re-emergence of epidemic malaria in the highlands of western Kenya. *Emerg Infect Dis* 1998; 4:671–676.

22. Mouchet JO, Manuin S, Sircoulon S, Laventure S, Faye, O, Onapa AW, Carnavale P, Julvez J, Fontenille D. Evolution of malaria for the past 40 years: Impact of climate and human factors. *J Am Mosq Control Assoc* 1998; 14:121–130.

23. Reiter P. Global warming and vector-borne disease. *Lancet* 1998; 351:839–840.

24. Epstein PR, Diaz HF, Elias S, et al. Biological and physical signs of climate change: Focus on mosquito-borne diseases. *Bull Am Meteorol Soc* 1998; 78:410–417.

25. Lindgren E, Tälleklint L, Polfeldt T. Impact of climatic change on the northern latitude limit and population density of the disease-transmitting European tick, *Ixodes ricinus*. *Environ Health Perspect* 2000; 108:119–123.

26. McMichael AJ, Haines A, Slooff R, Kovats S, editors. *Climate change and human health.* Geneva: World Health Organization, 1996.

27. McMichael AJ, Haines A. Global climate change: The potential effects on health. *BMJ* 1997; 315: 805–809.

28. Patz JA, Epstein PR, Burke TA, Balbus JM. Global climate change and emerging infectious diseases. *JAMA* 1996; 275:217–223.

29. Epstein PR. Health and climate. *Science* 1999; 285:347–348.

30. Epstein PR, editor. *Extreme weather events: the health and economic consequences of the 1997/98 El Niño and La Niña.* Cambridge, MA: Center for Health and the Global Environment, Harvard Medical School, 1999.

31. Parry ML, Rosenzweig C. Food supply and the risk of hunger. *Lancet* 1993; 342:1345–1347.

32. Bouma MJ, Sondorp HJ, van der Kaay HJ. Climate change and periodic epidemic malaria. *Lancet* 1994; 343:1440.

33. Bouma MJ, van der Kaay HJ. The El Niño southern oscillation and the historic malaria epidemics on the Indian subcontinent and Sri Lanka: An early warning system for future epidemics? *Trop Med Int Health* 1996; 1:86–96.

34. Epstein PR, Calix Pena OC, Racedo JB. Climate and disease in Colombia. *Lancet* 1995; 346:1243.

35. Bouma MJ, Dye C. Cycles of malaria associated with El Niño in Venezuela. *JAMA* 1997; 278: 1772–1774.

36. Nicholls N. El Niño southern oscillation and vector-borne disease. *Lancet* 1993; 342:1284–1285.

37. Hales S, Weinstein P, Woodward A. Dengue fever in the South Pacific: Driven by El Niño southern oscillation? *Lancet* 1996; 348:1664–1665.

38. Bouma MJ, Kovats S, Cox J, Goubet S, Haines A. A global assessment of El Niño's disaster burden. *Lancet* 1997; 350:1435–1438.

39. Kovats SJ, Bouma MJ, Haines A. *El Niño and health.* Geneva: World Health Organization, 1999.

40. Kalkstein LS. *The impact of weather and pollution on human mortality.* Washington: Office of Policy, Planning, and Evaluation, US Environmental Protection Agency, 1994. Report No.: EPA 230–94–019.

41. Kalkstein LS, Smoyer KE. The impact of climate change on human health: Some international applications. *Experientia* 1993; 49:469–479.

42. Smith and Lavender, Environmental Consultants; Sustainable Futures. *Climate Change Impacts: An Ontario Perspective.* Burlington, ON: Environment Canada, 1993.

43. Curwen M. Excess winter mortality: A British phenomenon? *Health Trends* 1991; 22:169–175.

44. Donaldson GC, Tchernjavskii VE, Ermakov SP, Bucher K, Keatinge WR. Winter mortality and cold stress in Yekaterinburg, Russia: Interview survey. *BMJ* 1998; 316:514–518.

45. Kalkstein LS. The impact of predicted climate change on human mortality. *Publications Climatol* 1988; 41:1–127.

46. Reeves WC, Hardy JL, Reisen WK, Milby MM. The potential effect of global warming on mosquito-borne arboviruses. *J Med Entomol* 1994; 31:323–332.

47. Epstein PR. Is global warming harmful to health? *Sci Am* 2000 Aug 283(2):50–57.

48. McMichael AJ. Integrated assessment of potential health impact of global environmental change: Prospects and limitations. *Environ Model Assess* 1997; 2:129–137.

49. Martens WJM, Niessen LW, Rotmans J, Jetten TH, McMichael AJ. Climate change and vector-borne disease: A global modeling perspective. *Global Environ Change* 1995; 5:195–209.

50. Martens WJM, Jetten TH, Focks DA. Sensitivity of malaria, schistosomiasis and dengue to global warming. *Climate Change* 1997; 35:145–156.

51. Patz JAS, Martens WJM, Focks DA, Jetten TH. Dengue fever epidemic potential as projected by general circulation models of global climate change. *Environ Health Perspect 1998;* 106:147–152.

52. Rogers DJ, Randolph SE. The global spread of malaria in a future, warmer world. *Science* 2000; 289:1763–1765.

53. Epstein PR, Ford TE, Colwell RR. Marine ecosystems. *Lancet* 1993; 342:1216–1219.

54. Harvell CD, Kim K, Burkholder JM, Colwell RR, Epstein PR, Grimes J, et al. Diseases in the ocean: Emerging pathogens, climate links, and anthropogenic factors. *Science* 1999; 285:1505–1510.

55. Valiela I. *Marine ecological processes.* New York: Springer-Verlag, 1984.

56. Hallegraeff GM. A review of harmful algal blooms and their apparent increase. *Phycologica* 1993; 32:79–99.

57. Colwell RR. Global climate and infectious disease: The cholera paradigm. *Science* 1996; 274:2025–2031.

58. MacKenzie WR, Hoxie NJ, Proctor ME, Gradus MS. A massive outbreak in Milwaukee of cryptosporidium infection transmitted through public water supply. *N Engl J Med* 1994; 331:161–167.

59. Duchin JS, Koster FT, Peters CJ, Simpson GL, Tempest B, Zaki SR, et al. Hantavirus pulmonary syndrome: A clinical description of 17 patients with a newly recognized disease. Hantavirus Study Group. *N Engl J Med* 1994; 330:949–955.

60. Epstein PR. Emerging diseases and ecosystem instability: New threats to public health. *Am J Public Health* 1995; 85:168–172.

61. Pini NC, Resa A, del Jesus Laime G, Lecot G, Ksiazek TG, Levis S, et al. Hantavirus infection in children in Argentina. *Emerg Infect Dis* 1998; 4:85–87.

62. Espinoza R, Vial P, Noriega LM, Johnson A, Nichol ST, Rollin PE, Wells R, et al. Hantavirus pulmonary syndrome in a Chilean patient with recent travel in Bolivia. *Emerg Infect Dis* 1998; 4:93–95.

63. Williams RJ, Bryan RT, Mills JN, Palma RE, Vera I, De Velasquez F, et al. An outbreak of hantavirus pulmonary syndrome in western Paraguay. *Am J Trop Med Hyg* 1997; 57(3):274–282.

64. Walker B, Steffen W, editors. *The terrestrial biosphere and global change: Implications for natural and managed ecosystems.* Stockholm: International Geosphere Biosphere Program, 1997.

65. Shriner DS, Street RB. North America. In: Watson RT, Zinyowera MC, Moss RH, editors. *The regional impacts of climate change: An assessment of vulnerability.* [A special report of the Intergovernmental Panel on Climate Change Working Group II.] New York: Cambridge University Press, 1998. pp. 253–330.

66. Rosenzweig C, Iglesias A, Yang XB, Epstein PR, Chivian E. Climate change and extreme weather events: Implications for food production, plant diseases, and pests. *Global Change and Human Health* 2001; 2:90–104.

67. Tilman D, Fargione J, Wolff B, D'Antonio C, Dobson A, Howarth R, et al. Forecasting agriculturally driven global environmental change. *Science* 2001; 292:281–284.

68. Environmental Data Services. *The unfinished climate business after Kyoto*. ENDS Report 1997; 275:16–20.

69. Working Group on Public Health and Fossil-Fuel Combustion. Short-term improvements in public health from global-climate policies on fossil-fuel combustion: An interim report. *Lancet* 1997; 350:1341–1349.

70. Wang X, Smith KR. *Near-term health benefits of greenhouse gas reductions: A proposed method of assessment and application to China*. Geneva: World Health Organization, 1998.

71. Haines A, Epstein PR, McMichael AJ. Global health watch: Monitoring impacts of environmental change. *Lancet* 1993; 342:1464–1469.

72. Leaf A. Potential health effects of global climate and environmental changes. *N Engl J Med 1989;* 21:1577–1583.

73. McMichael AJ. *Human frontiers, environments and disease: Past patterns, uncertain futures*. Cambridge: Cambridge University Press, 2001.

74. Epstein PR. Is global warming harmful to health? *Sci Am* 2000 Aug 283(2): 50–57.

Species Loss and Ecosystem Disruption

Eric Chivian

7

There is abundant evidence that human beings are beginning to alter, for the first time in our history, some of the planet's basic physical, chemical, and biological systems,[1–3] endangering other species and disrupting ecosystems in the process,[4] and ultimately threatening human health.[5–7]

The extinction of species is the most destructive and permanent of all assaults on the global environment, for it is the only one that is truly irreversible.[8] While species stocks may recover following mass-extinction events over the course of millions of years of evolution[9] and may fill in ecosystem niches that had once been occupied, it must be understood that these replacing species are new ones and that the unique information encoded by the DNA of the lost species is gone forever. This fact adds to the gravity of the current extinction crisis, the first that is the result of human activity.

Many authors have looked at the question of what will convince people to alter their behavior so as to preserve other species. Clearly spiritual, ethical, and aesthetic values, as well as economic factors, are powerful motivating forces. It may be argued that these considerations are not persuasive enough to result in any significant change in personal behaviors or in public policy, that the consequences of such global change may be too abstract and technical for most people to relate to. It may take an understanding of the direct, concrete, and personal risks to peoples' health and lives, and particularly to the health and lives of their children, for them to be motivated enough to protect the global environment. This insight has helped define the goals of this chapter: to summarize the potential human health consequences from species loss and the resultant disruption of ecosystems, to demonstrate that human health is inextricably linked to the health of the global environment, and to make clear a fundamental truth: that humans are "embedded in Nature."[10]

When *Homo sapiens* evolved some 120,000 years ago, the number of species on earth was the largest ever,[8] but human activity has resulted in species extinction rates that are currently 100 to 1,000 times those of pre-human levels.[11] While the record demonstrates that humans hunted to extinction scores of large mammals and birds as early as tens of thousands of years ago,[12,13] it is only in recent times that these extinctions have spread to virtually every part of the planet and to almost every phylum. Species numbers are now being reduced so rapidly that Ehrlich and Wilson and others have predicted that a quarter or more of all species now alive may become extinct during the next 50 years if these rates persist.[14,15] Such losses have prompted biologists to refer to the present period as "the sixth extinction."[16] (The last great extinction event, the fifth, was 65 million years ago, at the end of the Cretaceous Period when dinosaurs became extinct.)

Global climate change,[17] stratospheric ozone depletion,[18] chemical pollution,[19] acid rain,[20] the introduction of alien species,[21] and the over-hunting and overexploitation of species[22] all threaten biodiversity (the total complement of different living organisms in an environment), but the degradation, reduction, and fragmentation of habitats is currently the greatest threat, particularly in species-rich areas like tropical rain forests and coral reefs.[9,23]

Potential Medicines

With a loss of species, we are losing plants, animals, and microorganisms, perhaps most of which are still undiscovered, that may contain valuable new medicines. (Only about 1.5 million species have been identified,[24] but there are thought to be 10 or even 100 times that number.[11]) Over the course of millions of years of evolution, species have developed chemicals that have allowed them to fight infections, tumors, and other diseases; to capture prey and avoid being eaten; and to prevail over competitors in the struggle for territory. These chemicals that have become some of today's most important pharmaceuticals. Tropical rain forest organisms, for example, have given us D-tubocurarine (from the chondrodendron vine *Chondrodendron tomentosum),* synthetic analogues, indispensable for achieving deep-muscle relaxation during general surgery; quinine and quinidine (from the cinchona tree *Cinchona officianalis*), used in treating malaria and cardiac arrhythmias respectively (with quinine making a comeback as some of the infectious protozoans have developed resistance to new synthetic antimalarials); vinblastine and vincristine (from the rosy periwinkle plant *Vinca rosea*), which have revolutionized the treatment of leukemias and Hodgkin's lymphomas respectively; and erythromycin, neomycin, and amphotericin (from soil microbes of the genus *Streptomyces),* essential broad-spectrum antibiotics.

Temperate species have yielded some of our most useful drugs as well. Aspirin was synthesized from salicylic acid, originally extracted from the willow tree (*Salix alba*), and digitalis (from which digoxin and digitoxin were derived, used in treating congestive heart failure and atrial arrhythmias) came from the foxglove plant (*Digitalis purpurea).* Medications have also been developed from marine species, such as cytarabine from a Caribbean sponge *(Tethya crypta),* which is the single most effective agent for inducing remission in acute myelocytic leukemia.[25]

While some still argue that synthetic, combinatorial chemistry can produce all the medicines we need without relying on clues from nature (clues such as how natural toxins may provide important leads for the discovery of new medicines for neurological and cardiovascular diseases[26]), it is clear that plants, animals, and microbes have been, and will continue to be, critically important sources for biologically active compounds useful to medicine.[27] A recent survey of the 150 most frequently prescribed drugs in the United States, for example, has demonstrated that 57 percent of them contained compounds, or were patterned after compounds, derived from natural sources.[28]

Taxol and the Pacific Yew

The story of taxol and the Pacific Yew tree (*Taxus brevifolia*) illustrates how we may be foreclosing the possibility of finding new medicines, as species are lost before they have been analyzed for their chemical content. The commercially useless Pacific Yew was routinely discarded as a trash tree during logging of old-growth forests in the Pacific Northwest region of the United States until it was found by a National Cancer Institute natural-products screening program to contain the compound taxol, a substance that kills cancer cells by a mechanism unlike that of other chemotherapeutic agents known at the time—preventing cell division by inhibiting the disassembly of the mitotic spindle.[29] The discovery of the complex molecule taxol and its novel mechanism of action led to the synthesis of several taxol-like compounds that are even more effective than the natural compound,[30] illustrating how a clue from nature has led to the discovery of a new class of medicines that would have been extremely difficult, if not impossible, to discover in the lab. Early clinical trials have reported that taxol could induce remission in advanced ovarian cancer cases unresponsive to other treatments;[31] subsequent experience has shown that taxol may be one of the most promising new medicines available for breast and ovarian cancer.[30] (Taxol is the top-selling cancer drug in the United States.[32]) Other species with biomedical potential like the Pacific Yew are clearly being lost on a regular basis without our ever knowing what unique compounds they contain.

Conotoxins and Cone Snails

The diversity of peptide compounds in the venoms of cone snails, a genus of predatory snails (numbering about 500 species) inhabiting tropical coral-reef communities, is so great that it may rival that of alkaloids in higher plants and secondary metabolites in microorganisms.[33] Some of these compounds, which have been shown to block a wide variety of ion channels, receptors, and pumps in neuromuscular systems, have such selectivity that they have become important tools in neurophysiological research and may become invaluable to clinical medicine. One toxin has demonstrated great potency as an anticonvulsant.[34] Another, a voltage-sensitive calcium channel blocker, omega-conotoxin, binds with enormous specificity to neuronal calcium channels, and has shown potent activity in animals, both as an analgesic[35] and in keeping nerve cells alive following ischemia.[36] This compound is now in advanced clinical trials (in its synthetic form called SNX-111 or ziconotide) for two major clinical problems: (1) preventing nerve-cell death following coronary artery bypass surgery, head injury, and stroke; and (2) treating the chronic, intractable pain of patients with cancer, AIDS, and peripheral neuropathies.[37] SNX–111 has 1,000 times the analgesic potency of morphine, but, unlike morphine, it does not lead to a clouding of consciousness, to the loss of other peripheral sensations, or, of greatest importance, to the development of tolerance or addiction.[38] Although the several cone-snail species whose toxins have been studied have range sizes large enough so that their survival is not now threatened,[39] others, which are also likely to contain toxins of great medical importance, are at high risk because their ranges are small and because coral reefs are dying in many parts of the world.[40]

Research Models

Species loss also leads to the loss of valuable research models that help us understand human physiology and disease. Biomedical research has long relied on mice, rats, and guinea pigs as experimental subjects and on a host of other organisms possessing unique structures or physiologies—for example, fruit flies[41] and *E. coli* in genetics research; horseshoe crabs and squid for the study of nerve cells. Two groups of animals that are endangered, bears and sharks, and a microscopic roundworm that is not endangered provide examples of the kinds of biological secrets that could be lost to medical science with the loss of species.

Bears

Bear populations are threatened in many parts of the world because of the destruction of their habitats and because of overhunting secondary to the high prices their organs, reputed to have medicinal value, bring in Asian black markets.[42] Bear gallbladders, for example, are worth 18 times their weight in gold. There is also growing concern that greenhouse warming, by thinning Arctic sea ice, reducing seal populations, and altering when polar bears (*Ursus maritimus*—the largest land carnivore in the world) can feed, will threaten polar bear survival in the wild.[43,44]

Living bears are far more valuable than the sum of all their body parts. For example, in winter months black bears (*Ursus americanus*) enter a three- to seven-month period of hibernation called *denning,* in which they neither eat, drink, urinate, nor defecate, yet they can deliver young and nurse them, do not lose bone or lean body mass, and do not become ketotic or uremic[45] (polar bears have identical processes). Despite a lack of weight bearing for months, black bears do not suffer from osteoporosis, a condition that occurs during periods of immobility, or a lack of weight bearing, in all other mammalian species that have been studied, including humans.[46]

Understanding how bears prevent bone resorption could lead to new ways of preventing and treating osteoporosis, an overwhelming public health problem resulting in more than 1.5 million bone fractures and $13.8 billion in direct health care costs and lost productivity each year in the United States alone.[47] Similarly, learning how denning black bears avoid becoming uremic—because of their ability to convert urea into protein—could help in the development of effective treatments for end-stage renal disease,[48] which presently can be treated only by renal dialysis and kidney transplantation and for which public and private expenditures in the United States in 1998 came to almost $17 billion.[49]

Sharks

Why sharks, the first vertebrates to have developed antibodies and a full complement of immune proteins, may have a lower rate of infections and tumors and may be less likely to develop cancers when exposed to carcinogens, has prompted intensive investigation into the nature of their immune systems.[50–51] The answer may be that sharks produce powerful infection and cancer-fighting molecules in their tissues that boost their immune response, presumably the result of 450 million years of successful evolutionary experiments. One such substance isolated from sharks, an aminosteroid named squalamine, has prevented the growth of solid tumors in a number of animal models, most likely because of its angiogenesis-inhibiting ability.[52] Squalamine has also shown broad-spectrum antibiotic activity against Gram-negative and Gram-positive bacteria (with a potency comparable to ampicillin), as well as against some fungi and protozoa.[53] Because some shark species are endangered by overfishing and by the demand for shark-fin soup and for shark cartilage, our ability to understand their unique immune systems could be endangered as well.[54]

Elegans

With the mapping of the complete genome of the microscopic roundworm *C. elegans* (*Caenorhabditis elegans*), scientists have been able to design experiments to learn how the worm's genes control its metabolism of glucose,[55] the mechanisms of its programmed cell death,[56] and its longevity.[57,58] Though nematodes and mammals diverged from a common ancestor some 700 to 800 million years ago,[59] the genetic codes regulating these fundamental biological processes seem to have remained surprisingly similar for *C. elegans* and human beings. As a result, studying *C. elegans* may yield important new insights for understanding human diabetes; obesity; diseases in which there are disturbances in programmed cell death such as cancer, autoimmune disease, and neurodegenerative states; and aging.

Ecosystem Services

An ecosystem is made up of the sum total of all the species and their actions and interactions with each other and with the nonliving matter within a particular environment. How ecosystems provide the services that sustain all life on this planet remains one of the most complex and poorly understood areas of biological science.[60,61]

Ecosystem services include such vital functions as regulating the concentration of oxygen, carbon dioxide, and water vapor in the atmosphere; filtering pollutants from drinking water; regulating global temperature and precipitation; forming soil and keeping it fertile; pollinating plants; and providing food and fuel.[60,62,63] We cannot live without these life-support services, whose functioning can be disrupted by species loss.[4]

One critically important service is the role that biodiversity plays in controlling the emergence and spread of some infectious diseases in plants, animals, and humans. This protective function has only recently begun to be appreciated.[64–68] Examples of human infectious disease that can be affected by altering biodiversity include malaria and leishmaniasis through deforestation,[69] as well as Lyme disease by changes in the number of acorns and in the populations of black-legged ticks, white-footed mice, white-tailed deer, and hosts that are not competent for the transmission of the Lyme-disease bacteria.[70] Other examples include Argentine hemorrhagic fever by the replacement of natural grasslands with corn monoculture; [71] the outbreak of schistosomiasis in Lake Malawi through overfishing of the cichlid fish *Trematocranus plachydonan,* thought to be a predator of the snail intermediate host;[72,73] and cholera as a result of marine algal blooms, secondary in part to warming seas, coastal fertilizer runoff, and sewage discharge.[74]

Hantavirus Pulmonary Syndrome

The case of the hantavirus pulmonary syndrome outbreak in the American Southwest provides a valuable model about how altered numbers of a prey species carrying an infectious agent may result in the emergence of a lethal human infectious disease. Six years of drought in the Four Corners area of the American Southwest ended in the late winter and spring of 1993 with unusually heavy snow and rainfall. The drought, it was thought, had reduced the populations of the natural predators of the native deer mouse (owls, snakes, coyotes, and foxes), while the increased precipitation led to a very heavy crop of pine nuts and grasshoppers, food for the mice.[75] As a result, there was a twentyfold increase over baseline levels in the deer-mouse population.[76] At the same time, 17 previously healthy people, most of them Native Americans, developed a severe, rapidly progressive respiratory syndrome; 13 of them died.[77] It was subsequently learned that the deer mice carried a hantavirus that they shed in their saliva and excreta, and that, with the greater numbers of mice, there was a greater chance for people to come in contact with them and thus to become infected. In this case, a change in climate triggered the outbreak of a highly lethal infectious disease, but other alterations in species numbers, for example, through the elimination by overhunting of a major predator of a human infectious-disease carrier, might do so as well. It is not known how many viruses or other infectious agents in the environment, potentially harmful to humans, are being held in check by the natural equilibria afforded by biodiversity.

Prescriptions

Current high rates of species loss due to human activity call for a thorough reassessment of international species-protection efforts. By adding the lens of human health to this examination, we not only rightly include *Homo sapiens* as an integral part of the global environment, but we may also strengthen the motivation of our own species to protect others on which we depend. It is hoped that this unique perspective, largely missing at the present time from international deliberations on biodiversity, will help guide the formation of future policy agreements. Several policy implications result from this reassessment:

1. There is a need for a major international scientific effort that assesses the status of global biodiversity and ecosystem health on a regular basis, similar in scope and magnitude to the Intergovernmental Panel on Climate Change (IPCC). As with the IPCC, this assessment needs to include an examination of the contributions that plant, animal, and microbial species make to human health. Fortunately, both these initiatives are now underway. The Millennium Ecosystem Assessment (MEA), a project sponsored by the United Nations Environment Programme (UNEP), the United Nations Development Program (UNDP), the Global Environmental Facility, the World Health Organization, (WHO), the World Bank, and the World Resources Institute, will provide state-of-the art scientific input on global ecosystems to the United Nations every five years, beginning in 2003.[78] One chapter in the MEA report will be on human health, summarized from the full report of another project, Biodiversity: Its Importance to Human Health, will be written by an international team of scientists and led by the Center for Health and the Global Environment at Harvard Medical School under the auspices of the WHO and UNEP.[79] The biodiversity report will be presented to the UN in 2002.

2. Species that have been identified as having important biomedical potential need to be the subject of more focused protection efforts. For example, some cone-snail species are threatened by coral-reef degradation and by overcollection and trading. They should serve as models for more effective coral-reef protection efforts and should be included in CITES (Convention on International Trade in Endangered Species) protocols. Similarly, there needs to be far stricter enforcement of CITES, such as in the illegal trade of bear body parts on the Asian black market.

3. Much more attention needs to be paid to ecosystem services and to how certain critical species contribute to ecosystem functioning.

Significant increases in funding for research in this area are needed. We have only a primitive understanding of the mechanisms of most ecosystem services and of how they may be affected by environmental degradation.

4. The role of biodiversity in the emergence and spread of infectious diseases among plants, wildlife, and humans needs to be much better understood, along with how biodiversity loss may act synergistically with other forms of environmental change, including climate change, on infectious-disease dynamics.

5. Microbial and marine biodiversity are vastly underappreciated. As a result, we have almost no knowledge of the microbial and marine species that are being lost with environmental change or of their potential importance to medicine. As has been done for terrestrial ecosystems, aquatic and marine environments need to be analyzed for their "biodiversity hotspots" so that we can focus our conservation efforts in environments where they are most effective. International purchasing of these "hotspots" by governments and conservation organizations needs to accelerate greatly to protect them from development and degradation.

6. International mechanisms should be established to ensure that prospecting for medicines, by extensive sampling, does not endanger the species being studied, and that developing countries where such prospecting is taking place maintain control over their own natural resources and share in the profits of any commercial pharmaceuticals. Overharvesting of species for the traditional medicines, in addition, needs to be better controlled.

7. Greater collaboration among UNEP, the WHO, the Food and Agriculture Organization (FAO), UNDP, and the World Bank needs to be encouraged in dealing with the relationship of biodiversity and human health. It is noteworthy that UNEP and the WHO recently signed a Memorandum of Understanding on environmental health issues, which included a commitment to collaborate on biodiversity and human health matters.[80]

8. Finally, the main causes of species loss need to be addressed with a much greater sense of urgency—habitat destruction and fragmentation, climate change, invasive species, and overexploitation.

Conclusions

Despite an avowed reverence for life, human beings continue to destroy other species at an alarming rate, rivaling the great extinctions of the geological past. In the process, we are foreclosing the possibility of discovering the secrets they contain for the development of life-saving new medicines and of invaluable models for medical research, and we are beginning to disrupt the vital functioning of ecosystems, in ways we only vaguely comprehend, on which all life depends. We may also be losing some species so uniquely sensitive to environmental degradation that they may serve as our "canaries," warning us of future human health threats.

It may be that policymakers and the public will give protection of the global environment their highest priority only when they have begun to understand that human health and life ultimately depend on the health of other species and on the integrity of global ecosystems.

Physicians and other health professionals, veterinarians, conservation biologists, and other environmental scientists need to become knowledgeable about the human health dimensions of species loss and ecosystem disruption, for they may be the most powerful voices in helping to promote this understanding.

References

1. McCarthy JJ, Canziani OF, Leary NA, Dokken DJ, White KS, editors. Climate change 2001: Impacts, adaptation, and vulnerability. Cambridge, UK: Cambridge University Press, 2001.

2. Ozone Secretariat, United Nations Environment Programme. *Synthesis of the reports of the Scientific, Environmental Effects, and Technology and Economic Assessment Panels of the Montreal Protocol—1999, Nairobi.* Ozone Secretariat, United Nations Environment Programmes, 1999. Available: www.unep.ch/ozone

3. Vitousek PM, Mooney HA, Lubchenco J, Melillo JM. Human domination of earth's ecosystems. *Science* 1997; 277:494–499.

4. Chapin FS III, Walker BH, Hobbs RJ, et al. Biotic control over the functioning of ecosytems. *Science* 1997; 277:500–504.

5. McMichael AJ, Haines A, Slooff R, Kovats S, editors. *Climate change and human health.* Geneva: World Health Organization, 1996.

6. Longstreth J, De Gruijl FR, Kripke ML, Abseck S, Arnold F, Slaper HI, et al. Health risks [review]. *J Photochem Photobiol B* 1998; 46:20–39.

7. Chivian E. Environment and health: 7. Species loss and ecosystem disruption—The implications for human health. *CMAJ* 2001; 164:66–69.

8. Wilson EO. Threats to biodiversity. *Sci Am* 1989; 261(3):108–116.

9. Myers N, Mittermeier RA, Mittermeier CG, da Fonseca GA, Kent J. Biodiversity hotspots for conservation priorities. *Nature* 2000; 403:853–858.

10. Thomas L. The lives of a cell: Notes of a biology watcher. New York: Viking Press, 1974.

11. Pimm SL, Russell GJ, Gittleman JL, Brooks T. The future of biodiversity. *Science* 1995; 269:347–350.

12. Miller GH, Magee JW, Johnson BJ, Fogel ML, Spooner NA, McCulloch MT, et al. Pleistocene extinction of *Benyornis newtoni:* Human impact on Australian megafauna. *Science* 1999; 283:205–208.

13. Flannery TF. Paleontology: Debating extinction. *Science* 1999; 283:182–183.

14. Ehrlich PR, Wilson EO. Biodiversity studies: Science and policy. *Science* 1991; 253:758–762.

15. Baillie J, Groombridge T, editors. *1996 IUCN red list of threatened animals.* Gland, Switzerland: World Conservation Union, 1997.

16. Pimm SL, Brooks T. The sixth extinction: How large, where, and when? In: Raven PH, editor. *Nature and human society.* Washington, DC: National Academy of Sciences Press, 2000.

17. Pounds JA, Rodgen MPL, Campbell JH. Biological response to climate change on a tropical mountain. *Nature* 1999; 398:611–615.

18. Klesecker JM, Blaustein AR, Belden LK. Complex causes of amphibian population declines, *Nature* 2001:410:681–684.

19. Colburn T, vom Saal RS, Soto AM. Developmental effects of endocrine-disrupting chemicals in wildlife and humans. *Environ Health Perspect* 1993; 101:378–384.

20. Driscoll CT, Lawrence GB, Bulger AJ, Butler TJ, Cronan CS, Eager C. Acidic deposition in the Northeastern United States: Sources and inputs, ecosystem effects, and management strategies. *BioScience* 2001; 51(3):180–198.

21. Lovei GL. Global change through invasion. *Nature* 1997; 388:627–628.

22. Pauly D, Christensen V, Dalsgaard J, Froese R, Torres F. Fishing down marine food webs. *Science* 1998; 279:860–863.

23. Pimm SL, Raven R. *Extinction by numbers. Nature* 2000; 403:843–844.

24. May RM. How many species are there? *Science* 1988; 241:1441–1449.

25. Chabner BA, Allegra CJ, Curt GA, Calabresi P. Antineoplastic agents. In: Hardman JG, Limbird LE, Molinoff PB, Ruddon RW, editors. *Goodman and Gilman's the pharmacological basic of therapeutics.* 9th ed. New York: McGraw-Hill, 1996. p. 1251.

26. Harvey AL, Bradley KN, Cochran SA, Rowan EG, Pratt JA, Quillfeldt JA, et al. What can toxins tell us for drug discovery? *Toxicon* 1998; 36:1635–1640.

27. Newman DJ, Cragg GM, Snader KM. The influence of natural products upon drug discovery. *Nat Prod Rep* 2000:17:215–234.

28. Grifo R, Newman D, Fairfield AS, Bhattacharya B, Grupenhoff JT. The origins of prescription drugs. In: Grifo F, Rosenthal J, editors. *Biodiversity and human health*. Washington, DC: Island Press, 1997.

29. Cowden CJ, Paterson I. Cancer drugs better than taxol? *Nature* 1997; 387:238–239.

30. Nicolaou KC, Guy RK, Potier P. Taxoids: New weapons against cancer. *Sci Am* 1996; 274(6):94–98.

31. McGuire WP, Rowinsky EK, Rosenshein NB, Grumbine FC, Ettinger DS, Armstrong DK, et al. Taxol: A unique antineoplastic agent with significant activity in advanced ovarian epithelial neoplasms. *Ann Intern Med* 1989; 111:273–279.

32. Petersen M. Ivax says Bristol-Myers deal aims to delay a generic drug. *New York Times* 2000 Aug 16; Sect. C:6.

33. Olivera BM, Rivier J, Clark C, Ramilo CA, Corpuz GP, Abogadie FC, et al. Diversity of Conus neuropeptides. *Science* 1990; 249:257–263.

34. White HS, McCabe RT, Armstrong H, Donevan SD, Cruz LJ, Abogadie FC, et al. In vitro and in vivo characterization of conantokin-R, a selective NMDA antagonist isolated from the venom of the fish-hunting snail Conus radiatus. *J Pharmacol Exp Ther* 2000; 292:425–432.

35. Malmberg AB, Yaksh TL. Effects of continuous intrathecal infusion of omega-conopeptides on behavior and anti-nociception in the formalin and hot plate tests in rats. *Pain* 1994; 60:83–90.

36. Valentino K, Newcomb R, Gadbois T, Singh T, Bowersox S, Bitner S, et al. A selective N-type calcium channel antagonist protects against neuronal loss after global cerebral ischemia. *Proc Natl Acad Sci* 1993; 90:7894–7897.

37. Miljanich GP. Venom peptides as human pharmaceuticals. *Sci Med* 1997; 4(5):6–15.

38. Bowersox S, Tich N, Mayo M, Luther R. SNX–111: Antinociceptive agent N-type voltage-sensitive calcium channel blocker. *Drugs of the Future* 1998; 23(2):152–160.

39. Roberts C. Unpublished data. May 16, 2001.

40. Hayes RL, Goreau NI. The significance of emerging diseases in the tropical coral reef ecosystem. *Revista de Biologia Tropical* 1998; 46 Suppl 5:173–185.

41. Rubin GM, Lewis EB. A brief history of Drosophila's contributions to genome research. *Science* 2000; 287:2216–2218.

42. Montgomery S. Grisly trade imperils world's bears. *Boston Globe* 1992 Mar 2.

43. Stirling I, Derocher AE. Possible impacts of climatic warming on polar bears. *Arctic* 1993; 46:240–245.

44. Stirling I. The importance of polynyas, ice edges, and leads to marine mammals and birds. *J Marine Systems* 1997; 10:9–21.

45. Hasan Y, Morgan-Boyd RL, Dickson A, Meyer C, Nelson RA. Plasma carnitine levels in black bears. *Federation of American Societies of Exper Biol J* 1997; 11:A371.

46. Floyd T, Nelson RA, Wynne GF. Calcium and bone metabolic homeostasis in active and denning black bears. *Clin Orthop* 1990; 255:301–309.

47. National Osteoporosis Foundation. *Fast facts on osteoporosis.* Washington, DC: National Osteoporosis Foundation, 2000.

48. Nelson RA. Black bears and polar bears—Still metabolic marvels. *Mayo Clin Proc* 1987; 62:850–853.

49. U.S. Renal Data System. *USRDS 2000 annual data report.* Bethesda, MD: National Institute of Diabetes and Digestive and Kidney Diseases, National Institutes of Health (NIH), DHHS, 2000.

50. Litman GW. Sharks and the origins of vertebrate immunity. *Sci Am* 1996; 275(5):67–71.

51. Mestel R. Sharks' healing powers. *Nat Hist* 1996; 105(9):40–47.

52. Sills AK Jr, Williams JI, Tyler BM, Epstein DS, Sipos EP, Davis JD, et al. Squalamine inhibits angiogenesis and solid tumor growth in vivo and perturbs embryonic vasculature. *Cancer Res* 1998; 58:2784–2792.

53. Moore KS, Wehrli S, Roder H, Rogers M, Forrest JN Jr, McCrimmon D, et al. Squalamine: An aminosterol antibiotic from the shark. *Proc Natl Acad Sci* 1993; 90:1354–1358.

54. Malakoff D. Extinction on the high seas. *Science* 1997; 277:486–488.

55. Sze JY, Victor M, Loer C, Shi Y, Ruvkun G. Food and metabolic signalling defects in a *caenorhabditis elegans* serotonin-synthesis mutant. *Nature* 2000; 403:560–564.

56. Chen F, Hersh BM, Conradt B, Zhou Z, Riemer D, Gruenbaum Y, et al. Translocation of C. elegans CED–4 to nuclear membranes during programmed cell death. *Science* 2000; 287:1485–1489.

57. Kimura KD, Tissenbaum HA, Liu Y, Ruvkun G. daf-s, an insulin receptor-like gene that regulates longevity and diapause in Caenorhabditis elegans. *Science* 1997; 277:942–946.

58. Taub J, Lau JF, Ma C, Hahn JH, Hoque R, Rothblatt J, et al. A cytosolic catalase is needed to extend adult lifespan in C. elegans daf-C and clk-1 mutants. *Nature* 1999; 399:162–166.

59. Roush W. Worm longevity gene cloned. *Science* 1997: 277:897–898.

60. Daily GC, editor. *Nature's services: Societal dependence on natural ecosystems.* Washington, DC: Island Press, 1997.

61. Cohen JE, Tilman D. Biosphere 2 and biodiversity: The lessons so far. *Science* 1996; 274:1150–1151.

62. Costanza R, d'Arge R, de Groot R, Farber S, Grasso M, Hannon B, et al. The value of the world's ecosystem services and natural capital. *Nature* 1997; 387:253–260.

63. Aber JD, Melillo JM. Terrestrial ecosystems. 2nd ed. San Diego: Academic Press, 2001.

64. Zhu Y, Chen H, Fan J, Wang Y, Li Y, Chen J, et al. Genetic diversity and disease control in rice. *Nature* 2000; 406:718–722.

65. Wolfe MS. Crop strength through diversity. *Nature* 2000; 406:681–682.

66. Daszak P, Cunningham AA, Hyatt AD. Anthropogenic environmental change and the emergence of infectious diseases in wildlife. *Acta Tropica* 2001; 78:103–116.

67. Ostfeld RS, Keesing F. The function of biodiversity in the ecology of vector-borne zoonotic diseases. *Can J Zool* 2000; 78:2061–2078.

68. Epstein PR. Is global warming harmful to health? *Sci Am* 2000; 283(2):50–57.

69. Walsh JF, Molyneux DH, Birley MH. Deforestation: Effects on vector-borne disease. *Parasitology* 1993; 106:S55–S75.

70. Ostfeld RS, Keesing F. Biodiversity and disease risk: The case of Lyme disease. *Conserv Biol* 2000; 14(3):722–728.

71. Morse SS. Factors in the emergence of infectious diseases. *Emerg Infect Dis* 1995; 1(1):7–15.

72. Cetron MS, Chitsulo L, Sullivan JJ, Pilcher J, Wilson M, Noh J, et al. Schistosomiasis in Lake Malawi. *Lancet* 1996; 348:1274–1278.

73. Stauffer JR, Arnegard ME, Cetron M, Sullivan JJ, Chitsulo LA, Turner GF, et al. Controlling vectors and hosts of parasitic diseases using fishes. *BioScience* 1997; 47(1):41–49.

74. Colwell RR. Global climate and infectious disease: The cholera paradigm. *Science* 1996; 274:2025–2031.

75. Wenzel RP. A new hantavirus infection in North America. *N Engl J Med* 1994; 330:1004–1005.

76. Engelthaler DM, Mosley DG, Cheek JE, Levy CE, Komatsu KK, Ettestad P, et al. Climatic and environmental patterns associated with hantavirus pulmonary syndrome, Four Corners region, United States. *Emerg Infect Dis* 1999; 5(1):87–94.

77. Duchin JS, Koster FT, Peters CJ, Simpson GL, Tempest B, Zaki SR, et al. Hantavirus pulmonary syndrome: A clinical description of 17 patients with a newly recognized disease. *N Engl J Med* 1994; 330:949–955.

78. Ayensu E, Clasasen DR, Collins M, Dearing A, Fresco L, Gadgil M, et al. International ecosystem assessment. *Science* 1999; 286:685–686.

79. Biodiversity: Its importance to human health. See www.med.harvard.edu/chge/resources.html

80. World Health Organization, United Nations Environment Programme. Memorandum of understanding "enhancement of cooperation in the field of environmental health." Geneva: World Health Organization, 1999 Aug 23.

Ozone Depletion and Ultraviolet Radiation

Frank R. de Gruijl and Jan C. van der Leun

8

Ultraviolet (UV) radiation from the sun is responsible for a variety of familiar photochemical reactions, including photochemical smog, bleaching of paints, and decay of plastics. Conjugated bonds in organic molecules, such as proteins and DNA, absorb the UV radiation, which can damage these molecules. By a fortunate evolutionary event, the oxygen produced by photosynthesis forms a filter in the outer reaches of our atmosphere that absorbs the most energetic and harmful UV radiation, with wavelengths below 240 nm (in the UVC band: wavelength 100–280 nm). In the process, the oxygen molecules split up and recombine to form ozone (figure 8.1). In turn, this rarefied ozone layer (spread out between 10 and 50 km in the stratosphere but only 3 mm thick were it compressed at ground level) efficiently absorbs UV radiation of higher wavelengths (up to about 310 nm). A part of the UV radiation in the UVB band (wavelength 280–315 nm) still reaches ground level and is absorbed in sufficient amounts to have deleterious effects on cells. The less energetic radiation in the UVA band (wavelength 315–400 nm, bordering the visible band: wavelength 400–800 nm) is not absorbed by ozone and reaches ground level without much attenuation through a clear atmosphere (i.e., no clouds, no air pollution). Although not completely innocuous, the UVA radiation in sunlight is much less photochemically active and therefore generally less harmful than UVB radiation.

Life on earth has adapted itself to the UV stress, particularly UVB stress, by, for example, forming protective UV-absorbing surface layers,

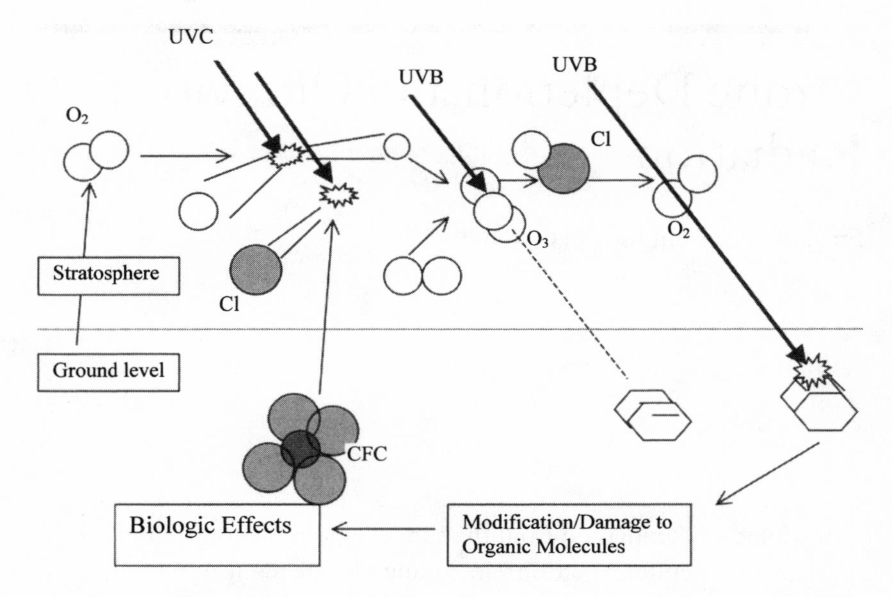

Figure 8.1 Simplified scheme of how the stratospheric ozone layer is formed and absorbs ultraviolet B (UVB) radiation and how chlorofluorocarbons (CFCs) degrade the ozone layer and increase UVB-induced biological effects. Molecular oxygen (O_2) and CFCs diffuse upward into the stratosphere, where they are decomposed by shortwave UV radiation (UVC). Atomic oxygen (O) recombines with molecular oxygen to form ozone (O_3). Ozone absorbs the short wavelengths in the UVB band, blocking photochemical reactions in organic molecules. Reactive chlorine (Cl) released from CFCs catalyzes the breakdown of ozone to molecular oxygen, which does not absorb UVB radiation. More UVB radiation thus passes through the stratosphere, enhancing photochemical reactions in organic molecules and increasing corresponding biological effects.

repairing cell damage, or entirely replacing damaged cells. Human skin shows all of these adaptive features. Our eyes are less well adapted, but they are shielded by the brows and by squinting.

Ozone Depletion

In 1974, Molina and Rowland,[1] who, together with Crutzen, received the 1995 Nobel Prize in Chemistry for their work in atmospheric chemistry, proposed that stratospheric ozone might be destroyed by industrially produced chlorine- and bromine-containing stable substances, such as chlorofluorocarbons commonly used in spray cans, refrigerators, and air-conditioners. These inert substances can reach the stratosphere, where they are decomposed by high-energy UV radiation and reactive chlorine and bromine are released (figure 8.1).

Although complicated by substantial daily, seasonal, and annual fluctuations (on the order of 20 percent), a gradual downward trend in the amount of stratospheric ozone has been measured in temperate and polar climate zones over the last two decades (e.g., 3 to 6 percent per decade at midlatitudes).[2,3] The exceptionally low levels of stratospheric ozone (over 50 percent depletion) that occur over Antarctica in the early austral spring were unanticipated.[4] This ozone "hole" develops because of the generation of reactive chlorine on the surface of ice crystals in high-altitude polar clouds.

The important consequence of a hole in the ozone layer is an increased flux of UV radiation at the earth's surface and in the top layers of the ocean. The productivity of phytoplankton was found to be reduced by 6 to 12 percent directly under the ozone hole.[5] Transient effects of increased UV radiation have thus been measured on organisms at the bottom of the marine food chain. What the long-term effects of ozone depletion will be on marine and terrestrial ecosystems is still uncertain.

Health Effects

Ozone depletion will not simply increase all UVB-related effects because adaptive reactions will dampen or even neutralize the impact of an increase in UVB radiation, even if sun-exposure behavior remains the same. For example, the UVB-driven production of vitamin D3 in the skin is self-limiting, and the serum level of the ultimately biologically active metabolite (1,25-dihydroxyvitamin D) is controlled by feedback mechanisms that maintain an approximately constant level.[6] Therefore, the beneficial effects from vitamin D3 are not expected to be enhanced in a large majority of people—the usual daily sun exposure of head and hands is already sufficient. Among people with regular sun exposure (not sunbathers), rates of sunburn are not likely to increase either because the human skin can adapt very efficiently to gradual increases in UVB exposure. Photodermatoses may even become less frequent. The loss of adaptation to UV radiation during winter generally increases the incidence and severity of photodermatoses.

Because the ozone depletion will be strongest in wintertime, this loss of adaptation will decrease.

The health effects believed to be affected by an increased flux in UV radiation include skin cancer, cataracts, and cellular immunity. The decreased immunity may cause increased severity of infections, less effective vaccinations, and perhaps an increase in non-Hodgkin's lymphoma.

Skin Cancer

UVB radiation is known to damage DNA at specific sites (neighboring pyrimidines). Mutations in the p53 tumor suppressor gene in human skin carcinomas were found at precisely these sites.[7,8] (The p53 protein plays an important role in the cellular response to DNA damage: cell cycle arrest to allow more time for DNA repair, and apoptosis if the cell is overly damaged. Its dysfunction can cause genomic instability and thus lead to the dysfunction of other genes.) These point mutations in the p53 gene resembled those found in experiments with UV-irradiated cell cultures (mainly a cytosine replaced by a thymine, and sometimes even two neighboring cytosines replaced by two thymines).[7,8] The high frequency and apparent selection of the p53 mutations in the tumors appears to constitute the most direct evidence that UVB radiation contributes to the formation of skin cancer. The risk of squamous cell carcinoma appears to increase with lifelong sun exposure, whereas accumulated exposure does not appear to contribute to the risk of basal cell carcinoma.[9] The risk of basal cell carcinoma[10] and cutaneous malignant melanoma[11] appears to be related to childhood exposure and to intermittent overexposures. Skin cancer is by far the most common form of cancer among white people in the United States, and the incidence rates of all three types of skin cancer increase with increasing ambient UVB radiation loads, although each type of cancer has a different dose–response curve.[12,13] Surprisingly, however, experiments in fish[14] and opossums[15] have indicated that UVA radiation may be more important than UVB radiation in the induction of melanoma.

The effective carcinogenic UV dose of a certain spectrum of sunlight (which varies with solar angle and ozone thickness) can be ascertained by adding up the effective doses contributed by different wavelengths. The wavelength dependence of squamous cell carcinoma induction in humans can be estimated from data obtained in mouse experiments,[16,17] and it crudely parallels the wavelength dependence of sunburn (peaking between 290 and 300 nm). Thus, forecasting UV indices to warn against sunburn should help to minimize carcinogenic UV exposures. In general, people need daily exposure to ordinary levels of UVB radiation to maintain their vitamin D levels and their adaptation to sun exposure, but overexposure will contribute to adverse effects, such as immunosuppression and skin cancer.

The relation between ambient UVB levels and skin cancer–incidence rates at different geographic locations can be used to estimate the impact of ozone depletion on skin cancer rates.[18] However, it is not well established whether melanoma induction in humans is dominated by UVB radiation; it may be dominated by UVA radiation, as explained earlier. It is therefore uncertain how melanoma will be affected by ozone depletion, which will change only the UVB component of sunlight. Because melanomas constitute a small proportion (<10 percent) of skin cancers, they will not greatly affect skin cancer–incidence projections. Melanomas, however, are the most lethal type of skin cancer, and, if their incidence is not affected by ozone depletion, the projected increases in skin cancer mortality will be overestimated by 25 to 50 percent.

Based on the international agreements to phase out ozone-depleting substances, as described in the Montreal Protocol of 1987 and its later amendments,[19] the thickness of the stratospheric ozone layer is expected to reach a minimum around the turn of the century, after which the incidence of skin cancer is projected to reach a maximum (10 percent higher than current rates) around 2060 and then fall.[18] However, this projection assumes full compliance of all nations and industries with the international agreements. Unrestrained emissions of chlorofluorocarbons and the use of methylbromide in agriculture would be expected to produce a gradual and persistent degradation of the ozone layer and likely lead to a dramatic increase in skin cancer incidence, which we have estimated could quadruple by the year 2100.[18]

Cataracts

The effects of ozone depletion and UV radiation on skin cancer are well studied. Other health effects are less certain, mainly because the wavelength and UV dose dependencies are not adequately established. By assuming that the UVB dominance found in cataract formation in animal experiments also holds for cataract development in humans, we have estimated that the incidence of cataracts would ultimately rise by 0.5 percent for every 1 percent persistent decrease in ozone.[20] Dolin reviewed epidemiological studies and concluded that there is good evidence for a relation between (cortical) cataracts and sunlight.[21]

Immune Effects

Most of the data on immunomodulatory effects of UV radiation stem from mouse experiments, but the suppression of contact hypersensitivity by UV light also occurs in humans, along with many of the cellular effects (disappearance of Langerhans cells, influx of macrophage-like cells, and release of certain cytokines such as interleukin–1, tumor necrosis factor α, and interleukin–10).[22,23] UV radiation can alter organic molecules in the skin and thus form novel antigens ("neoantigens"), which could potentially evoke immune reactions. The UV-induced immunosuppression could therefore serve to prevent these illicit immune reactions (such as occur in photodermatoses). However, this immunosuppresion can clearly have adverse effects. UV effects on viral, bacterial, parasitic, and fungal infections have been established in animal experiments.[24,25] Infections that have a phase in the skin, like malaria and leprosy, are likely to be influenced. The herpes simplex "cold sore" triggered by sun exposure is a good example of an infection in humans brought forth by UV-induced immunosuppression.[26] Reports by Finsen more than 100 years ago showed that UV irradiation can influence infections both beneficially (skin tuberculosis or lupus vulgaris) and adversely (smallpox).[27]

Immunosuppression is a risk factor for both skin cancer and non-Hodgkin's lymphoma. UV-induced immunosuppression might be expected to play a role in the prevalence of these cancers.[28] The risk of non-Hodgkin's lymphoma is associated with skin cancer, and the incidence rates of both types of cancer have risen significantly over the past decades. This does not necessarily imply that increased sun exposure is the common causative factor. Unlike skin cancer, non-Hodgkin's lymphoma does not show a clear latitudinal gradient over the United States.[29] More recently, however, it has been reported that the incidence of non-

Hodgkin's lymphoma in Sweden tends to increase toward the south, but to a considerably lesser extent than that of squamous cell carcinoma.[30] Experimental data showed that a combination of a chemical carcinogen and UV exposure induced lymphomas in hairless mice, whereas each agent on its own did not.[31]

Indirect Health Effects

There is the possibility that human health would not only be directly affected by the increased levels of ambient UVB radiation, but also indirectly affected by changes to the environment because of stratospheric ozone depletion. For example, UVB radiation contributes to photochemical smog. It can increase ozone at ground level and thus aggravate respiratory afflictions. A potentially more dramatic effect could stem from a decrease in food production because of the effects of stratospheric ozone depletion on certain plants and animals. These indirect health effects are, however, even less easily quantified than the direct effects on human health.

Prescriptions

Overall, compliance with the international agreements on phasing out ozone-depleting substances has been good, but there is no reason for complacency. As the production of the bulk of such substances in developed countries is curbed, other sources like methylbromide become more important. These other sources may prove more difficult to phase out. For example, the production and replacement of ozone-depleting substances can be difficult to control in some developing countries because there is no suitable infrastructure or proper channeling of international support to overcome the locally prohibitive costs for the required changes.[2] Moreover, there is always the danger that certain countries that have or have not ratified the agreements will continue to produce or increase production of ozone-depleting substances. To discourage any subversive production of such substances, consumers in developed countries should remain vigilant and not purchase anything operating or produced with ozone-depleting substances when good alternatives are available.

Despite a long-standing concern about the effects of ozone depletion, our grasp of the problem is still incomplete and limited to only a few effects. Our knowledge of the effects of current levels of UV exposure on ecosytems and human health is scanty, which hampers any extrapolation to effects at higher levels of UV exposure. There is an obvious need to remedy the situation, if only to weigh benefits against costs of the next human activity that may threaten the ozone layer.

Acknowledgments

We thank the Dutch Ministry of Housing, Zonal Planning and Environment for financially supporting our work for the UNEP Panel on the Environmental Effects of an Ozone Depletion. We also thank the European Commission and the Netherlands Cancer Foundation for subsidizing our research.

References

1. Molina MJ, Rowland FS. Stratospheric sink for chlorofluoromethanes: Chlorine atom-catalysed destruction of ozone. *Nature* 1974; 249:810–812.

2. Ozone Secretariat, United Nations Environment Programme. *Synthesis of the reports of the Scientific, Environmental Effects, and Technology and Economic Assessment Panels of the Montreal Protocol—1999.* Nairobi: Ozone Secretariat, United Nations Environment—Programme, 1999. Available: www.unep.org/ozone and www.unep.ch/ozone

3. McKenzie R, Connor B, Bodeker G. Increased summertime UV radiation in New Zealand in response to ozone loss. *Science* 1999; 285:1709–1711.

4. Farman JC, Gardiner BG, Shanklin JD. Large losses of total ozone in Antarctica reveal seasonal ClO_x NO_x interaction. *Nature* 1985; 315:207–210.

5. Smith RC, Prezelin BB, Baker KS, Bidigare RR, Boucher NP, Coley T, et al. Ozone depletion: Ultraviolet radiation and phytoplankton biology in Antarctic waters. *Science* 1992; 255:893–1040.

6. Holick MF, MacLaughlin JAM, Parrish JA, Anderson RR. The photochemistry and photobiology of vitamin D3. In: Regan JD, Parrish JA, editors. *The science of photomedicine.* New York: Plenum Press, 1982. pp. 195–218.

7. Brash DE, Rudolph JA, Simon JA, Lin A, McKenna GJ, Baden HP, et al. A role for sunlight in skin cancer: UV-induced p53 mutations in squamous cell carcinoma. *Proc Natl Acad Sci USA* 1991; 88:10124–10128.

8. Ziegler A, Leffell DJ, Kunala S, Sharma HW, Gailani M, Simon JA, et al. Mutation hotspots due to sunlight in the p53 gene of nonmelanoma skin cancers. *Proc Natl Acad Sci USA* 1993; 90:4216–4220.

9. Vitasa BC, Taylor HR, Strickland PJ, Rosenthal FS, West S, Abbey H, et al. Association of nonmelanoma skin cancer and actinic keratosis with cumulative solar ultraviolet exposure in Maryland watermen. *Cancer* 1990; 65:2811–2817.

10. Kricker A, Armstrong BK, English DR, Heenan PJ. Does intermittent sun exposure cause basal cell carcinoma? A case–control study in Western Australia. *Int J Cancer* 1995; 60:489–494.

11. Holman CDJ, Armstrong BK. Cutaneous malignant melanoma and indicators of total accumulated exposure to the sun: An analysis separating histogenic types. *J Natl Cancer Inst* 1984; 73:75–82.

12. Scotto J, Fears TR. *Incidence of nonmelanoma skin cancer in the United States.* Washington, DC: US Department of Health and Human Services, 1981. Cat. No. NIH 82–2433.

13. Scotto J, Fears TR. The association of solar ultraviolet and skin melanoma incidence among Caucasians in the United States. *Cancer Invest* 1987; 5:275–283.

14. Setlow RB, Grist E, Thompson K, Woodhead AP. Wavelengths effective in induction of malignant melanoma. *Proc Natl Acad Sci USA* 1993; 90:6666–6705.

15. Ley RD. Ultraviolet radiation A-induced precursors to cutaneous melanoma in Monodelphis domestica. *Cancer Res* 1997; 57:3682–3684.

16. De Gruijl FR, Sterenborg HJCM, Forbes PD, Davies RE, Kelfkens G, Van Weelden H, et al. Wavelength dependence of skin cancer induction by ultraviolet irradiation of albino hairless mice. *Cancer Res* 1993; 53:53–60.

17. De Gruijl FR, Van der Leun JC. Estimate of the wavelength dependency of ultraviolet carcinogenesis in humans and its relevance to the risk assessment of a stratospheric ozone depletion. *Health Phys* 1994; 67:319–325.

18. Slaper H, Velders GJM, Daniel JS, De Gruijl FR, Van der Leun JC. Estimates of ozone depletion and skin cancer incidence to examine the Vienna Convention achievements. *Nature* 1996; 384:256–258.

19. Ozone Secretariat, United Nations Environment Programme. *The 1987 Montreal protocol on substances that deplete the ozone layer [as adjusted and amended by the second, fourth, seventh and ninth meetings of the parties].* Nairobi: Ozone Secretariat, United Nations Environment Programme, 1997. Available: www.unep.ch/ozone/mont_t.htm

20. Van der Leun JC, De Gruijl FR. Influence of ozone depletion on human health. In: Tevini M, editor. *UV-B radiation and ozone depletion.* Boca Raton, FL: Lewis Publishers, 1993. pp. 95–123.

21. Dolin PJ. Ultraviolet radiation and cataract: A review of epidemiological evidence. *Br J Ophthalmol* 1994; 78:178–182.

22. Yosikawa T, Rae V, Bruin-Slot W, Van den Berg W, Taylor JR, Streilein JW. Susceptibility to effects of UV-B radiation on the induction of contact hypersensitivity as a risk factor for skin cancer in humans. *J Invest Dermatol* 1990; 95:530–536.

23. Cooper KD, Oberhelman L, Hamilton TA, Baardsgaard O, Terhune M, LeVee G, et al. UV exposure reduces immunization rates and promotes tolerance to epicutaneous antigens in humans: Relationship to dose, CD1a-DR+ epidermal macrophage induction and Langerhans cell deletion. *Proc Natl Acad Sci USA* 1992; 89:8497–8501.

24. Jeevan A, Brown E, Kripke ML. UV and infectious diseases. In: Krutmann J, Elmets CA, editors. *Photoimmunology*. Oxford: Blackwell Science, 1995. pp. 153–163.

25. Goettsch W, Garssen J, De Gruijl FR, Van Loveren H. Effects of UV-B on the resistance against infectious diseases. *Toxicol Lett* 1994; 72:259–263.

26. Miura S, Kulka M, Smith CC, Imafuku S, Burnett JW, Aurelian L. Cutaneous ultraviolet radiation inhibits herpes simplex virus–induced lymphoproliferation in latently infected subjects. *Clin Immunol Immunopathol* 1994; 72:62–69.

27. Finsen NR. Über die Bedeutung der chemische Strahlen des Lichtes für Medicin und Biologie. Leipzig: Vogel Publishing, 1899.

28. Goldberg LH. Basal-cell carcinoma as predictor for other cancers. *Lancet* 1997; 349:664–665.

29. Hartge P, Devesa SS, Grauman D, Fears TR, Fraumeni JF Jr. Non-Hodgkin's lymphoma and sunlight. *J Natl Cancer Inst* 1996; 88:298–300.

30. Adami J, Gridley G, Nyren O, Dosemeci M, Linet M, Glimelius B, et al. Sunlight and non-Hodgkin's lymphoma: A population-based cohort study in Sweden. *Int J Cancer* 1999; 80:641–645.

31. Husain Z, Pathak MA, Flotte T, Wick MM. Role of ultraviolet radiation in the induction of melanocytic tumors in hairless mice following 7,12-dimethylbenz(a)antracene application and ultraviolet irradiation. *Cancer Res* 1991; 51:4964–4970.

Additional Reading

De Gruijl FR. Health effects from solar UV radiation. *Radiat Protect Dosimetry* 1997; 72:177–196.

Longstreth J, De Gruijl FR, Kripke ML, Abseck S, Arnold F, Slaper HI, et al. Health risks. *J Photochem Photobiol B* 1998; 46:20–39.

Madronich S, McKenzie RL, Björn LO, Caldwell MM. Changes in biologically active ultraviolet radiation reaching the earth's surface. *J Photochem Photobiol B* 1998; 46:5–19.

Related Web Sites

European Ozone Research Coordinating Unit: www.ozone-sec.ch.cam.ac.uk
Impacts of a projected ozone depletion: www.gcrio.org/CONSEQUENCES/summer95/impacts.html
Ozone Depletion, US Environmental Protection Agency: www.epa.gov/ozone/index.html
Ozone Secretariat, United Nations Environment Programme: www.unep.org/ozone
Stratospheric Ozone and Human Health Project, Center for International Earth Science Information Network: sedac.ciesin.org/ozone
US Global Change Research Information Office: www.gcrio.org

Environmental Endocrine Disruption

Gina M. Solomon and Ted Schettler

9

During the past 50 years, tens of thousands of synthetic chemicals have been synthesized and released into the general environment. Some of these chemicals unintentionally interfere with hormone function in animals and, in some cases, in humans. The public health implications of these so-called endocrine disruptors have been the subject of scientific debate, media interest, and policy attention over the past several years. The current scientific debate centers on whether there is evidence of health effects or significant risks to the general human population from exposure to these chemicals.[1]

A wave of interdisciplinary research over the past few years has demonstrated that chemicals in our environment can interfere with endocrine function. Although in adults the endocrine system exhibits the ability to recover from fairly significant perturbations, in the fetus even minor changes in hormone levels can result in lifelong effects.

Endocrine disruption is increasingly a subject of the popular press and of justifiable public concern. Many public policy decisions today involve chemicals known or suspected to interfere with hormone function, so input from the health care community and the public on this issue is extremely important. In this chapter, we review the history of environmental endocrine disruption, the mechanisms of action of endocrine disruptors, and current evidence of effects on reproduction, infant development, and neurobehavioral function. Finally, we discuss health policy activities worldwide relevant to endocrine-disrupting chemicals in the environment.

Historical Background

Endocrine disruption is not a newly discovered phenomenon. In the 1930s, studies in laboratory animals demonstrated estrogenic properties of a number of industrial chemicals, including bisphenol A, now widely used in plastics, resins, and dental sealants. The feminizing effect of the pesticide DDT in roosters was reported in the 1950s.[2]

Although hormonally active chemicals are widely used for beneficial medical purposes, adverse effects have also occurred. In 1971, clinicians traced an epidemic of vaginal clear cell carcinoma in young women to maternal use of a synthetic estrogen, diethylstilbestrol (DES), during pregnancy. DES daughters have an increased risk of reproductive and immunological abnormalities, while sons are at risk of genital anomalies and abnormal spermatogenesis.[3] In animals, and possibly in humans, DES alters male- and female-typical behavior patterns.[4] The example of DES indicates that the fetus, rather than the adult, may be most at risk from the adverse effects of hormonal disruption.

Mechanisms of Action and Fetal Vulnerability

It is known that some pesticides and other industrial chemicals can directly bind to, or block, hormone receptors, thereby initiating or blocking receptor-activated gene transcription—the production of proteins from genetic information.[5] Other environmental chemicals act indirectly on hormonal balance by altering hormone production, hormone transport on binding proteins, receptor numbers on target organs, or hormone metabolism.[6] For example, polychlorinated biphenyls (PCBs) interfere with thyroid function by a variety of mechanisms, including increased metabolism of the thyroid hormone T_4, interference with T_4 delivery to the developing brain by displacement from the carrier protein, and interference with the conversion of T_4 to the active form of thyroid hormone known as T_3.[7]

During development, the fetus is particularly sensitive to hormonal fluctuations. Low-level exposures to hormones or toxicants may result in permanent physiological changes not seen in adults exposed at similar levels.[8] For example, mild hypothyroidism in an adult is not expected to have long-term effects on the brain. In contrast, subtle hypothyroidism during fetal and neonatal life causes disruption of neurotransmitters, nerve growth factors, nerve-cell growth, and normal energy production in the developing brain, altering cognitive and neuromotor development.[9,10]

Potential Health Implications

Reported abnormalities in laboratory animals and wildlife exposed to endocrine-disrupting chemicals include feminization of males, abnormal sexual behavior, birth defects, altered sex ratio, lower sperm density, decreased testis size, altered time to puberty, cancers of the mammary glands or testis, reproductive failure, and thyroid dysfunction (see table 9.1).[11–13]

Table 9.1 Examples of endocrine-disrupting chemicals

Chemical	Use	Mechanism	Health effect	Reference
DES	Synthetic estrogen	Direct binding to estrogen receptor	Prenatal exposure: females—vaginal cancer, reproductive tract abnormalities; males—cryptorchidism, hypospadias, semen abnormalities	3
Methoxychlor	Insecticide	Metabolite binds directly to the estrogen receptor	Rodent—accelerated vaginal opening, abnormal estrus; aggressive behavior (male)	57, 57
DDT	Insecticide	Metabolite (DDE) binds directly to the androgen receptor	Rodent (male)—delayed puberty, reduced sex accessory gland size, altered sex differentiation	57
Vinclozolin	Fungicide	Binds directly to the androgen receptor	Rodent (male)—reduced anogenital distance, nipple development, hypospadias	57
Polychlorinated biphenyls (PCBs)	No longer manufactured; still found in electrical transformers, capacitors, toxic waste sites, food chain	Accelerated T_4 metabolism; decreased T_4, elevated thyroid stimulating hormone (TSH) (high doses—thyromimetic)	Human (in utero exposure)—delayed neurological development; IQ deficits,	57, 57

Table 9.1 (continued)

Chemical	Use	Mechanism	Health effect	Reference
Atrazine	Herbicide	Reduces gonadotropin releasing hormone (GnRH) from the hypothalamus; reduces pituitary luteinizing hormone (LH) levels; interferes with the metabolism of estradiol, blocks estrogen receptor binding; increases aromatase-related conversion of androgens to estrogens	Rodents—mammary tumors in female rats, alters estrus cycling Humans—some evidence of breast, ovarian, and prostate tumors	57, 57, 57, 57, 57, 57
di-2-ethylhexyl-phthalate (DEHP)	Plasticizer for poly-vinylchloride plastics, including medical products	Decreases testosterone synthesis during sexual differentiation	Rodents—in utero exposure: reduced testis weight and ano-genital distance	57
Dioxin	By-product of industrial processes including waste incineration; food contaminant	Binds to Ah (aryl hydrocarbon) receptor. Increases estrogen metabolism; decreases estrogen-mediated gene transcription, decreases estrogen levels; decreases testosterone levels by interfering with hypothalamic-pituitary-gonadal (HPG) axis	Rodents—in utero exposure: Female —delayed vaginal opening, increased susceptibility to mammary cancer Male—decreased testosterone, hypospadias, hypospermia, delayed testicular descent; feminized sexual behavior Human—decreased T_3, T_4; decreased testosterone;* cancer*	57, 57, 57, 57,57

*adult exposures

Epidemiologic studies have found associations between exposure to specific pesticides or industrial chemicals and thyroid stimulating hormone (TSH), testosterone, and prolactin levels in adults.[14–16] Some of these studies have also found significant associations with other relevant endpoints including diminished sperm quality, impaired sexual function, and testicular cancer.[17,18] Numerous studies have found associations between occupational solvent or pesticide exposure and subfertility or adverse effects on offspring, but it is not clear whether these are due to endocrine mechanisms.[19] Several case-control studies have shown increased risks of cryptorchidism (undescended testicles) and hypospadias (a birth defect of the penis) among the sons of farmers and of women gardeners exposed to pesticides.[20] These findings are important because animal studies have shown the same effects after exposure to estrogenic or antiandrogenic chemicals.

Population-based epidemiologic studies relevant to endocrine disruption are few. The studies that have been done are weakened by the inevitable time lag between exposure and clinical disease, difficulty defining exposed and control populations, and poor retrospective assessment of exposures during the prenatal period. Moreover, limited understanding of the role of gene-environment interactions increases the likelihood that susceptible subpopulations may remain unidentified. Perhaps as a result, epidemiologic data concerning the relationship between breast cancer and tissue levels of certain organochlorines, such as DDT, its by-product DDE, PCBs, or dieldrin, are conflicting.[21,22]

Surveillance-based studies in the general population show increases in some potentially hormone-related conditions (table 9.2). These increases are not completely explained by improved detection or reporting.[23] Although behavioral and nutritional factors are potential explanations for some of these observations, it is biologically plausible, and consistent with laboratory and wildlife evidence, that fetal exposure to endocrine-disrupting chemicals may be playing a role. Cancers and other health effects may manifest many years later as steroid hormones continue to stimulate cell growth and proliferation.[24–26]

Hormones and Neurobehavioral Effects

Exposure to certain environmental pollutants has been associated with learning and behavioral abnormalities following prenatal or early postnatal

Table 9.2 Trends in human diseases potentially related to endocrine function

Endpoint	Region	Trend	Degree of change	Reference
Hypospadias (birth defect of the penis)	Canada U.S.	Increasing	4.3% per year 3.3% per year	57
Cryptorchidism (undescended testicles)	Canada U.S.	Increasing	3.5% per year 1.6% per year	44
Sperm count	Canada U.S. Europe	Decreasing	−0.7%/year* −1.5% /year −3% / year	57 57 46
Testicular cancer	Canada U.S. Europe	Increasing	2.1% per year 2.3% per year 2.3%–5.2% per year†	57 57 57
Prostate cancer	Canada U.S.	Increasing	3% per year‡ 5.3% per year	57 57
Breast cancer	Saskatchewan U.S.	Increasing	3.3% per year 1.9% per year	57 57
Sex ratio	Canada U.S.	Shift toward females	−1.0 males/ 10,000/year −0.5 males/ 10,000/year	57
Age at breast development	U.S.	Shifting earlier	11.2–9.96 years in whites	57

*When data from prior to 1984 are included this trend disappears.
†Range is dependent on country, with Sweden at the lower and East Germany at the upper end of the range.
‡International trends in prostate cancer are complicated by the introduction of the prostate specific antigen (PSA) screen, but prostate cancer mortality also increased (U.S. and Canadian mortality increased by about 1% per year through 1995), implying that improved diagnosis may not fully explain the rising incidence trends.

exposures. In some cases there is evidence that these neurological abnormalities may be due to an endocrine mechanism. For example, a single low dose of dioxin during development of the hypothalamic-pituitary-gonadal (HPG) axis in the rat produces a feminizing effect on the behavior of male offspring, reflecting altered sexual differentiation of the brain.[27]

Thyroid hormone is well known to affect development of the fetal brain.[10] Thyroid disruption, including goiter and neurobehavioral abnormalities, is found in wildlife and laboratory animal populations feeding on the organochlorine-contaminated food chain of the Great Lakes.[28–30] In utero and lactational exposure of nonhuman primates to environmentally relevant levels of PCBs causes impaired learning.[31] In humans, higher levels of PCBs in breast milk correlate with elevated TSH levels in nursing infants.[32] Blood levels of certain PCBs positively correlate with TSH levels and negatively correlate with free T_4 levels in children aged 7–10.[33] Many researchers therefore believe that PCBs may affect brain development by suppressing thyroid hormone during brain development.[34] In addition to the antithyroid effects of PCBs, animal studies also reveal evidence of altered neurotransmitter and neuroreceptor levels, which may be primary or secondary to the thyroid effects.[35]

Studies of children exposed to PCBs prenatally through maternal consumption of Great Lakes fish reveal delayed psychomotor development and increased distractibility in those most highly exposed.[36,37] In one study, at age 11 the most exposed children were more than three times as likely to perform poorly on IQ tests and more than twice as likely to be at least two years behind in reading comprehension.[38] Current body burdens of PCBs in certain groups, such as Inuit children in Canada, exceed levels known to adversely affect cognitive functioning.[39]

Beyond Endocrine Disruption: Other Signaling Pathways

Although much attention has been focused on the endocrine system, disruption of other biological signaling pathways is an important related issue. The structural and functional development of the brain is dependent on integration of the actions of hormones, neurotransmitters, neurotrophins, and locally produced steroids.[40]

Some chemicals, such as organophosphate pesticides, interfere with neurotransmitter levels and binding. The acute toxic effects are mostly reversible in the adult. In the developing brain, however, effects are more likely to be permanent. Most of the relevant data are based on animal studies. For example, rodents exposed to a single low dose of an organophosphate pesticide in a critical period of neonatal life have permanently decreased brain density of one type of nerve receptor (muscarinic receptors) and hyperactive behavior when tested as adults.[41] Similar effects are seen following neonatal challenge with DDT and pyrethroid insecticides.[42] Recent research has demonstrated that in the developing brain, neurotransmitters perform growth-regulatory functions.[43] For example, acetylcholine influences cell division, differentiation, [44] and synapse formation.[45] Inhibition of acetylcholinesterase alters these processes, as well as reducing neurite outgrowth in individual neurons.[46] The precise mechanisms responsible for these findings are not yet well understood, but it is clear that immature neuroendocrine systems are sensitive to low doses of exogenous agents that have no apparent effect on adults.

Policy Implications

The topic of endocrine disruption has brought to the surface an underlying debate about the nature of scientific proof and decisions about whether to take action in the face of scientific uncertainty. Some argue that there is no conclusive proof of human health effects due to endocrine disruption at current exposure levels in the general population. Others point to suggestive evidence and warn that the consequences of inaction may be significant for future generations.

Detecting health effects due to exposure to endocrine-disrupting chemicals is difficult and will continue to pose substantial challenges. Hormones act at extremely low (part-per-trillion) levels; therefore, even low-level exposures to hormonally active agents may be of concern, particularly during sensitive periods of fetal development when protective feedback loops have not yet developed. Endocrine-mediated effects may be subtle and are often delayed, making it difficult to link early life exposures to illnesses or conditions that may only become apparent in adulthood. Moreover, adverse impacts may be manifest primarily in populations rather than individuals. For example, slight overall declines in sperm density or intelligence may have little relevance for an individual but important adverse implications for the population.[47] Surveillance systems that focus on individuals rather than populations may, therefore, fail to detect early evidence of impacts until substantial damage has already occurred.

Low-level exposures to endocrine-disrupting chemicals are ubiquitous in today's environment. Persistent chemicals such as DDT, PCBs, and dioxins are detectable in nearly 100 percent of human blood samples, and even some of the shorter-lived potential endocrine disruptors, such as phthalates and other (non-DDT) organochlorine pesticides, are detected in general-population surveys of blood or urinary residues.[48,49]

Despite substantial evidence of the impacts of endocrine-disrupting chemicals in humans and wildlife, there is still considerable uncertainty about the severity and scope of the health threat. The ubiquitous nature of exposures, nontrivial potential health effects, and the difficulties inherent in quantifying those effects justify mitigating exposures now, while the remaining scientific uncertainties are being addressed. This is the conclusion of the International Joint Commission with respect to Great Lakes water quality, and it serves as an instructive example of how to establish policy under conditions of scientific uncertainty.

The International Joint Commission and the POPs Treaty

The International Joint Commission (IJC), created by a treaty between the United States and Canada in 1909 to prevent disputes along the border and in the Great Lakes, has taken a leadership role in defining a "persistent toxic substance" and targeting such chemicals for elimination. Many of the chemicals of concern to the IJC are also endocrine disruptors, and some are still in commercial use today. In 1992, in its *Sixth Biennial Report Under the Great Lakes Water Quality Agreement of 1978,* the IJC advocated a weight-of-evidence approach to the identification of substances that may cause harm.[50] The commission linked persistent toxic substances, particularly organochlorines, in the Great Lakes to injury, disease, and death in a variety of life forms, including humans. The Science Advisory Board (SAB) of the IJC found that the class of chlorinated organics tends to exhibit persistence and toxicity, and that evaluating and regulating these substances on a chemical-by-chemical basis is impractical and unscientific.[51]

The SAB recommended, therefore, that organochlorines be treated as a class and be subject to phaseout, except in individual cases in which the weight of evidence supports the view that a chemical does not threaten health and the integrity of ecosystems. Subsequently, the IJC recommended that all affected parties develop timetables to sunset the use of chlorine and chlorine-containing compounds as industrial feedstocks and that means to reduce or eliminate other uses be examined.

These recommendations embody a framework that includes zero discharge from human activities, consideration of the full life cycle of products and processes, the weight of evidence, and the reversal of the burden of proof. This framework recognizes that the effects of chemicals on health and ecosystems is complex, that cause-and-effect relationships are often difficult to establish with certainty, but that incontrovertible proof of causal links between individual chemicals and individual effects is not a reasonable standard since the substances exist as mixtures in the environment and produce complex effects.[51] In this case, the concept of "reverse onus" shifts the burden of proof to the manufacturer or user of a chemical to demonstrate that the evidence suggests safety rather than risks of harm, in keeping with a precautionary approach. (See chapter 14.)

In 1994, in its *Seventh Biennial Report Under the Great Lakes Water Quality Agreement of 1978,* the IJC said, "It is generally agreed, in principle, that the burden of proof concerning the safety of chemicals should lie with the proponent for the manufacture, import, or use of at least substances new to commerce in Canada and the United States, rather than with society as a whole to provide absolute proof of adverse impacts."[52] Then, in 1998, in its *Ninth Biennial Report,* the IJC stated, "Injury has occurred in the past, is occurring today and, unless society acts now to further reduce the concentration of persistent toxic substances in the environment, injury will continue in the future. The fact that such injury is occurring, coupled with a lack of knowledge about other, as yet unrecognized effects is a call for action by all [Great Lakes] basin stakeholders to minimize and eliminate injury."[53]

Here the IJC has laid out an approach intended to protect human health and the environment from harm, while recognizing the limits of science and acknowledging the difficulties inherent in reducing scientific uncertainty in complex ecosystems. The important conclusion is that incontrovertible proof of harm is not and should not be necessary before taking precautionary action. This conclusion is directly relevant to endocrine-disrupting chemicals because of the nature of their health effects and the difficulty establishing causal relationships with single chemicals in complex ecosystems.

The global effort to phase out 12 persistent organic pollutants (POPs), including DDT, PCBs, dieldrin, dioxin, and other known or suspected endocrine-disrupting chemicals, has resulted in a treaty signed in Stockholm in May 2001. The Stockholm Convention is an important part of the effort to reduce human exposures to endocrine-disrupting chemicals. The treaty will move beyond a phaseout of the 12 POPs listed and will employ a precautionary approach to identify and take action against numerous other chemicals that are environmentally persistent, bioaccumulate, and may pose a health risk to future generations. The treaty establishes a scientific POPs Review Committee to evaluate additional chemicals for inclusion in the treaty. Most important, the treaty specifies that "lack of full scientific certainty shall not prevent" a chemical from being included in the provisions of the treaty.[54]

Screening, Testing, and Tracking

Further research into exposures and effects of endocrine disruptors is essential as well as education and preventive action to reduce exposures. The U.S. Environmental Protection Agency and the Organization for Economic Cooperation and Development (OECD) are developing screening and testing programs to evaluate the effects of a variety of environmental chemicals on estrogen, androgen, and thyroid hormone function.[55,56] Research programs on basic mechanisms, wildlife effects, exposures, and human health implications are ongoing throughout the world.[57,58]

Improved monitoring of disease and exposure is essential for tracking trends in subtle, delayed effects of environmental exposures. Birth-defect registries, using active case ascertainment rather than passive reporting, can provide important data on structural birth defects. Neurodevelopmental abnormalities are exceedingly difficult to monitor, yet the evidence suggests that further investigation of time trends and causes is urgently needed. Ongoing efforts around the world to develop accurate biological monitoring of blood and urine for numerous chemical toxicants will improve exposure assessment in epidemiologic studies and may eventually provide tools for physicians to assess risks to individuals. As these tests become standardized and widely available, they will be useful clinically, just as blood-lead testing has helped facilitate a range of interventions that have resulted in a marked reduction of lead poisoning.

A science-based precautionary approach is a traditional part of medicine and public health. Public health practitioners have historically worked to prevent adverse health effects through education and practical exposure reduction whenever feasible.[59] In the case of endocrine disruptors, such an approach will require the medical community to play a critical role in evaluating the science and educating the public. The public has a right to know about the scientific evidence and the uncertainties surrounding the health effects of environmental chemicals. Endocrine-disrupting chemicals pose a risk to wildlife and humans that may manifest gradually and as a subtle decrement in normal functioning. These changes may be irreversible if precautionary measures are not taken. Because many endocrine disruptors are environmentally persistent and can travel great distances in air and water currents, an international perspective is critical to success. International approaches embodied in the International Joint Commission and in the Stockholm Convention provide a foundation for scientifically based phaseout of hazardous chemicals.

Acknowledgments

The authors gratefully acknowledge the support of the W. Alton Jones Foundation.

References

1. Committee on Hormonally Active Agents in the Environment, National Research Council. *Hormonally active agents in the environment.* Washington, DC: National Academy Press, 1999.

2. Burlington H, Lindeman VF. Effect of DDT on testes and secondary sex characteristics of white leghorn cockerels. *Proc Soc Exp Biol Med* 1950; 74:48–51.

3. Giusti RM, Iwamoto K, Hatch EE. Diethylstilbestrol revisited: A review of the long-term health effects. *Ann Int Med* 1995; 122:778–788.

4. Reinisch JM, Ziemba-Davis M, Sanders SA. Hormonal contributions to sexually dimorphic behavior in humans. *Psychoneuroendocrinology* 1991; 16:213–278.

5. Cooper RL, Kavlock RJ. Endocrine disruptors and reproductive development: A weight-of-evidence overview. *J Endocrinol* 1997; 152:159–166.

6. Bradlow HL, Davis DL, Lin G, Sepkovic D, Tiwari R. Effects of pesticides on the ratio of 16a/2-hydroxyestrone: A biologic marker of breast cancer risk. *Environ Health Perspect* 1995; 103 Supp l7:147–150.

7. Brouwer A, Morse DC, Lans MC, Schuur AG, Murk AJ, Klasson-Wehler E, et al. Interactions of persistent environmental organohalogens with the thyroid hormone system: Mechanisms and possible consequences for animal and human health. *Toxicol Ind Health* 1998; 14:59–84.

8. Bigsby R, Chapin RE, Daston GP, Davis BJ, Gorski J, Gray LE, et al. Evaluating the effects of endocrine disruptors on endocrine function during development. *Environ Health Perspect* 1999; 107 Suppl 4:613–618.

9. den Ouden AL, Kok JH, Verkerk PH, Brand R, Verloove-Vanhorick SP. The relation between neonatal thyroxine levels and neurodevelopmental outcome at age 5 and 9 years in a national cohort of very preterm and/or very low birth weight infants. *Pediatr Res* 1996; 39:142–145.

10. Haddow JE, Palomaki GE, Allan WC, Williams JR, Knight GJ, Gagnon J, et al. Maternal thyroid deficiency during pregnancy and subsequent neuropsychological development of the child. *New Engl J Med* 1999; 341:549–555.

11. Crisp TM, Clegg ED, Cooper RL, Wood WP, Anderson DG, Baetcke KP, et al. Environmental endocrine disruption: An effects assessment and analysis. *Environ Health Perspect* 1998; 106 Suppl 1:11–56.

12. vom Saal FS, Cooke PS, Buchanan DL, Palanza P, Thayer KA, Nagel SC, et al. A physiologically based approach to the study of bisphenol A and other estrogenic chemicals on the size of reproductive organs, daily sperm production, and behavior. *Toxicol Ind Health*. 1998; 14:239–260.

13. McLachlan JA, Newbold RR, Li S, Negishi M. Are estrogens carcinogenic during development of the testes? *APMIS* 1998; 106:240–242.

14. Steenland K, Cedillo L, Tucker J, Hines C, Sorensen K, Deddens J, et al. Thyroid hormones and cytogenetic outcomes in backpack sprayers using ethylenebis(dithiocarbamate) (EBDC) fungicides in Mexico. *Environ Health Perspect* 1997; 105:1126–1130.

15. Sweeney MH, Calvert GM, Egeland GA, Fingerhut MA, Halperin WE, Piacitelli LA. Review and update of the results of the NIOSH medical study of workers exposed to chemicals contaminated with 2,3,7,8-tetrachlorodibenzodioxin. *Terat Carcinog Mutagen* 1997–98; 17:241–247.

16. Manzo L, Artigas F, Martinez E, Mutti A, Bergamaschi E, Nicotera P, et al. Biochemical markers of neurotoxicity: A review of mechanistic studies and applications. *Hum Exp Toxicol* 1996; 15: Suppl 1:S20–S35.

17. Hardell L, Ohlson CG, Fredrikson M. Occupational exposure to polyvinyl chloride as a risk factor for testicular cancer evaluated in a case-control study. *Int J Cancer* 1997; 73:828–830.

18. Whelan EA, Grajewski B, Wild DK, Schnorr TM, Alderfer R. Evaluation of reproductive function among men occupationally exposed to a stilbene derivative: II. Perceived libido and potency. *Am J Ind Med* 1996; 29:59–65.

19. Gold EB, Tomich E. Occupational hazards to fertility and pregnancy outcome. *Occup Med* 1994; 9:435–469.

20. Weidner IS, Moller H, Jensen TK, Skakkebaek NE. Cryptorchidism and hypospadias in sons of gardeners and farmers. *Environ Health Perspect* 1998; 106:793–796.

21. Hunter DJ, Hankinson SE, Laden F, Colditz GA, Manson JE, Willett WC, et al. Plasma organochlorine levels and the risk of breast cancer. *N Engl J Med* 1997; 337:1253–1258.

22. Hoyer AP, Grandjean P, Jorgensen T, Brock JW, Hartvig HB. Organochlorine exposure and risk of breast cancer. *Lancet* 1998; 352:1816–1820.

23. Toppari J, Larsen JC, Christiansen P, Giwercman A, Grandjean P, Guillette LJ, et al. Male reproductive health and environmental xenoestrogens. *Environ Health Perspect* 1996; 104 Suppl 4:741–803.

24. Sharpe RM, Skakkebaek NE. Are oestrogens involved in falling sperm counts and disorders of the male reproductive tract? *Lancet* 1993; 341:1392–1395.

25. Gardner WA. Hypothesis: The prenatal origins of prostate cancer. *Hum Path* 1995; 26:1291–1292.

26. Davis DL, Axelrod D, Osborne M, Telang N, Bradlow HL, Sittner E. Avoidable causes of breast cancer: The known, unknown, and the suspected. *Ann NY Acad Sci* 1997; 833:112–128.

27. Mably TA, Moore RW, Goy RW, Peterson RE. In utero and lactational exposure of male rats to 2,3,7,8-tetrachlordibenzo-p-dioxin. 2. Effects on sexual behavior and the regulation of LH secretion in adulthood. *Toxicol Appl Pharmacol* 1992; 114:108–117.

28. Leatherland JF. Changes in thyroid hormone economy following consumption of environmentally contaminated Great Lakes fish. *Toxicol Ind Health* 1998:14:41–57.

29. Daly HB, Stewart PW, Lunkenheimer L, Sargent D. Maternal consumption of Lake Ontario salmon in rats produces behavioral changes in the offspring. *Toxicol Ind Health* 1998; 14:25–39.

30. Moccia RD, Fox GA, Britton A. A quantitative assessment of thyroid histopathology of herring gulls from the Great Lakes and a hypothesis on the causal role of environmental contaminants. *J Wildlife Dis* 1986; 22:60–70.

31. Rice DC. Behavioral impairment produced by low-level postnatal PCB exposure in monkeys. *Environ Res* 1999; 80:S113–S121.

32. Koopman-Esseboom C, Morse DC, Weisglas-Kuperus N, Lutkeschipholt IJ, Van der Paauw CG, Tuinstra LG, et al. Effects of dioxins and polychlorinated biphenyls on thyroid hormone status of pregnant women and their infants. *Pediatr Res* 1994; 36:468–473.

33. Osius N, Karmaus W, Kruse H, Witten J. Exposure to polychlorinated biphenyls and levels of thyroid hormones in children. *Environ Health Perspect* 1999; 107:843–849.

34. Porterfield SP. Thyroidal dysfunction and environmental chemicals—potential impact on brain development. *Environ Health Perspect* 2000; 108 Suppl 3:433–438.

35. Kodavanti PR, Tilson HA. Neurochemical effects of environmental chemicals: In vitro and in vivo correlations on second messenger pathways. *Ann NY Acad Sci* 2000; 919:97–105.

36. Jacobson JL, Jacobson SW. Effects of in utero exposure to PCBs and related contaminants on cognitive functioning in young children. *J Pediatrics* 1990; 116; 38–45.

37. Loney E, Reihman J, Darvill T, Mather J, Daly H. Neonatal behavioral assessment scale performance in humans influenced by maternal consumption of environmentally contaminated Lake Ontario fish. *J Great Lakes Res* 1996; 22:198–212.

38. Jacobson JL, Jacobson SW. Intellectual impairment in children exposed to polychlorinated biphenyls in utero. *N Engl J Med* 1996; 335:783–789.

39. Muckle G, Dewailly E, Ayotte P. Prenatal exposure of Canadian children to polychlorinated biphenyls and mercury. *Can J Public Health* 1998; 89 Suppl 1:S20–S25, S22–S27.

40. Lauder JM. Neurotransmitters as morphogens. *Prog Brain Res* 1988; 73:365–387.

41. Ahlbom J, Fredriksson A, Eriksson P. Exposure to an organophosphate (DFP) during a defined period in neonatal life induces permanent changes in brain muscarinic receptors and behavior in adult mice. *Brain Res* 1995; 677:13–19.

42. Eriksson P, Johansson U, Ahlbom J, Fredriksson A. Neonatal exposure to DDT induces increased susceptibility to pyrethroid (bioallethrin) exposure at adult age—Changes in cholinergic muscarinic receptor and behavioural variables. *Toxicology* 1993; 77:21–30.

43. Lauder JM, Schambra UB. Morphogenic roles of acetylcholine. *Environ Health Perspect* 1999; 107 Suppl 1:65–69.

44. Slotkin TA. Developmental cholinotoxicants: Nicotine and chlorpyrifos. *Environ Health Perspect* 1999; 107 Suppl 1:71–80.

45. Das KP, Barone S Jr. Neuronal differentiation in PC12 cells is inhibited by chlorpyrifos and its metabolites: Is acetylcholinesterase inhibition the site of action? *Toxicol Appl Pharmacol* 1999; 160:217–230.

46. Bigbee JW, Sharma KV, Gupta JJ, Dupree JL. Morphogenic role for acetylcholinesterase in axonal outgrowth during neural development. *Environ Health Perspect* 1999; 107 Suppl 1:81–87.

47. Rice DC. Issues in developmental neurotoxicology: interpretation and implications of the data. *Can J Public Health* 1998; 89 Suppl 1:S31–S36, S34–S40.

48. Blount BC, Silva MJ, Caudill SP, Needham LL, Pirkle JL, Sampson EJ, et al. Levels of seven urinary phthalate metabolites in a human reference population. *Environ Health Perspect* 2000; 108:979–982.

49. Archibeque-Engle SL, Tessari JD, Winn DT, Keefe TJ, Nett TM, Zheng T. Comparison of organochlorine pesticide and polychlorinated biphenyl residues in human breast adipose tissue and serum. *J Toxicol Environ Health* 1997; 52:285–293.

50. International Joint Commission. Sixth Biennial Report Under the Great Lakes Water Quality Agreement of 1978. Washington, DC and Ottawa, ON: 1992.

51. Muir T, Eder P, Muldoon P, Lerner S. *A strategy for virtual elimination of persistent toxic substances.* Vol 2, appendices. Washington, DC, and Ottawa, ON: International Joint Commission, 1993.

52. International Joint Commission. Seventh Biennial Report Under the Great Lakes Water Quality Agreement of 1978. Washington, DC and Ottawa, ON: 1994.

53. Legault L. *Ninth biennial report.* Washington, DC, and Ottawa, ON: 1998. pp. 11.

54. See http://www.ipen.org

55. See US EPA Web site at http://www.epa.gov/scipoly/oscpendo/

56. See OECD Web site at http://www.oecd.org/ehs/endocrin.htm

57. Reiter LW, DeRosa C, Kavlock RJ, Lucier G, Mac MJ, Melillo J, et al. The U.S. federal framework for research on endocrine disruptors and an analysis of research programs supported during fiscal year 1996. *Environ Health Perspect* 1998; 106:105–113.

58. See, for example, US EPA Web site at http://www.epa.gov/endocrine/

59. Horton R. The new public health of risk and radical engagement. *Lancet* 1998; 352:251–252.

Body Burdens of Industrial Chemicals in the General Population

Joe Thornton, Michael McCally, and Jeff Howard

10

The second industrial revolution spawned an immense chemical industry that provides materials now used in virtually every sector of commerce. Petroleum and other materials are transformed by industrial processes into fuels, plastic, pesticides, cosmetics, food additives, and pharmaceuticals. In the United States, some 70,000 individual industrial chemicals are registered with the Environmental Protection Agency for commercial use and are sold in the marketplace, some in billions of pounds per year.[1] Only a very small fraction of these substances have been characterized for potential biological activity or human toxicity,[2] and approximately 1,500 new chemicals are introduced into the market each year.[3] Thousands more novel compounds are generated as by-products of industrial processes or as degradation products and metabolites of other synthetic chemicals; even fewer toxicological data are available for these substances.

Residues of human-made substances can now be found in the air, soil, water, and food web in the most remote reaches of the planet.[4–6] Pollutants that are distributed ubiquitously result in universal human exposure through inhalation, drinking water, and the food supply. Some of the substances to which the general human population is exposed resist metabolism and excretion and therefore accumulate in body tissues. The quantity of an exogenous substance or its metabolites that has accumulated in an individual or population is defined as a *body burden*.

In this chapter, we present information about the levels of industrial chemicals, organochlorines in particular, in the tissues and fluids of the general population, and we interpret the significance of these measurements. We focus on the class of organochlorines because there is ample data on their presence in human tissues and considerable concern about their global accumulation, toxicity, and impact on human and wildlife health. We discuss the basic pharmacokinetic principles behind pollutant bioaccumulation to clarify the limits of body-burden measurements, and we review the technical difficulties of past and current analytic methods. We also explore the implications of these measurements for risk communication and policymaking in pollution prevention. Since the toxicology of organochlorines and other industrial chemicals has been described in other recent publications,[4–10] we limit our discussion to the potential impacts on public health of the universal chemical body burden. We draw on some ideas and materials from a recent book by the first author, with permission of the publisher.[4]

Body-Burden Estimation

An individual's body burden of a pollutant is estimated by measuring the concentration of that substance in one or more tissues, usually by gas chromatography/mass spectrometry (GC/MS).[10,11] Chemical body burdens are complex and dynamic in a number of ways, and these characteristics make a full characterization of the general public's body burden exceedingly difficult. First, the body burden of a pollutant is not stable over time. It reflects a dynamic balance between the amount taken in and the amount excreted or metabolized into another material.[12] Many industrial chemicals, like formaldehyde, benzene, and some pesticides, are taken into the body but are broken down and rapidly excreted or metabolized, producing a negligible long-term body burden, although levels after an acute exposure may be high. Chemicals that are persistent are those that resist metabolic alteration and excretion and/or are tightly bound to the tissues in which they are stored.

Second, body burdens are not distributed homogeneously within an individual: the partitioning of a pollutant among various tissues and fluids reflects the substance's degradability and affinity for fats, minerals, and other endogenous materials. Metabolic models have been developed for a number of toxic chemicals.[13] These models summarize available knowledge about the metabolism of a specific chemical and predict the distribution of the chemical in the body and the rate at which it will be eliminated. A general model showing possible metabolic pathways of a xenobiotic compound and body materials that can be used for biological monitoring was prepared by a working group of the National Research Council[14] (figure 10.1). Compartments may be organs (e.g., liver and brain), fluids (blood or extracellular fluid), or matrices (adipose tissue or bone). The choice of compartment in biomonitoring for any pollutant will affect the level measured, the limit of detection, and the recency of the exposure being estimated.

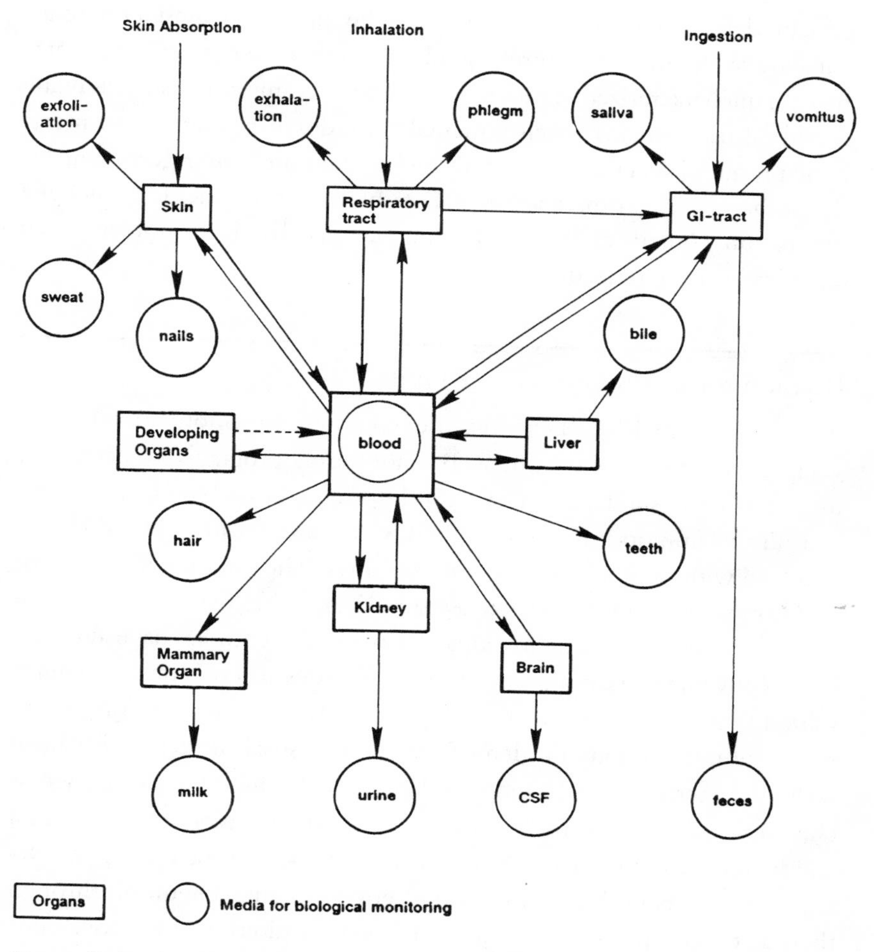

Figure 10.1 A general model showing the possible metabolic pathways of an exogenous or xenobiotic compound by three routes of exposure and their association with target tissue dose. (CSF = cerebral spinal fluid) (reference 14)

Third, the body burden of an individual in today's environment consists of hundreds of synthetic substances. The ability of chemical analyses based on GC/MS to characterize the full range of contaminants is limited in a number of ways.[12] Compounds can be identified only if they are present in concentrations above a detection limit (usually in parts per trillion or billion). Substances present in very low quantities will not be detected, but if there are hundreds or thousands of them, together they may make up the bulk of the total chemical burden. Moreover, routine analyses can identify only compounds that can be matched against a reference database of chemical signatures, so substances that are not yet in the database go uncharacterized. Thus, novel or exotic compounds, such as many industrial by-products, environmental breakdown products, and metabolic products, will remain unidentified in even the most rigorous analysis. It has been estimated that the fatty tissues of the U.S. general population contain at least 700 contaminants that have not yet been chemically characterized.[15]

Biomonitoring Programs in the United States

Public health officials and scientists use biomonitoring information for surveillance, control, and treatment. Biomonitoring programs, to date, have measured toxic substances in specific human tissues—blood, serum, fat, hair, urine, amniotic and seminal fluid, saliva, and breath—in individuals with and without unusual chemical exposures. The purpose of many biomonitoring programs has been to assess the health risks of occupationally or environmentally exposed individuals. Some of these studies have also analyzed convenience samples of "normal" persons to establish a reference value against which to compare highly exposed or otherwise at-risk groups.

The human body burden of specific industrial substances has been well characterized in selected populations with high exposures, such as chemical accident victims,[16,17] agricultural workers,[18] workers in chemical and incineration industries,[19,20] military personnel exposed to Agent Orange,[21] and persons highly exposed to environmental chemicals through their diet[22,23] or specific contaminated food.[24] In addition, numerous studies have sought to identify specific compounds in samples drawn from individuals in the general population with no known special exposures. Reports in this category, including many of those cited in table 10.1, have

Table 10.1 Organochlorine chemicals in the tissues and fluids of the general North American population

Substance	Present in	Reference
Aliphatic organochlorines		
Bromochlorotrifluoroethane	adipose	11
Bromochlorotrifluoroethane	breath	19
Carbochlorodithioic acid, S-methyl ester	adipose	11
Carbon tetrachloride	milk, blood, breath	8, 13, 7, 17, 14
Chlordecane	milk, breath	8, 19
Chloroethane	milk	8
Chloroform	adipose, milk, blood, breath	1, 8, 13, 7, 17, 14
Chlorohexane	milk, breath	8, 17
Chloromethane	milk, breath	8, 19
Chloropentane	milk	8
Chlorotrifluoromethane	milk	8
1,1-dichloro-l-propene	adipose	11
1,4-dichlorobutane	adipose	11
Dibromochloromethane	milk	8
Dibromodichloromethane	milk	8
Dichloroacerylene	breath	17
Dichlorodifluoromethane	milk, breath	8,7
Dichlorofluoromethane	breath	19
1,2-dichloroethane	breath	7
1,1-dichloroethylene (vinylidene chloride)	adipose	11
Dichloroethylene	milk	8
Dichloromethane	milk, blood, breath	8, 13, 7
1,1-dichloro-l-nitromethane	blood	13
Dichloronitromethane	breath	17
Dichloropropene	milk	8
Trichlorotrifluoroethane	milk, breath	8,7
Hexachloro-1,3-buradiene	adipose	18
1,1,1,2-tetrachloroethane	breath	14
1,1,2,2-tetrachloroethane	adipose, breath	1,14
Tetrachloroethylene	adipose, milk, blood, breath	1, 8, 13, 7, 17, 14

Table 10.1 (continued)

Substance	Present in	Reference
1,1,1-trichloroethane	adipose, milk, breath	1, 8, 7, 17, 14
1,1,2,-trichloroethane	breath	7
Trichloroethylene	adipose, milk, breath	15, 8, 7, 14
Trichlorofluoromethane	breath	7
tris(2-chloroethyi)phosphate	adipose	1
tris (dichloropropy 1) phosphate	semen	10
Chlorinated aromatics		
Chlorobenzene	adipose, milk, breath	1, 8, 7, 14
1,2-dichlorobenzene	adipose, milk, breath	1, 25, 7, 32
1,3-dichlorobenzene	adipose, milk, breath	11, 25, 14
1,4-dichlorobenzene	adipose, milk, breath	1, 25, 17
Clichlorobenzene (congener unidentified)	adipose, blood	1, 13
1,2,3-trichlorobenzene	adipose, milk	32, 25
1,2,4-trichlorobenzene	adipose, milk	1, 25, 32
1,3,5-trichlorobenzene	milk	25
1,2,5-trichlorobenzene	adipose	32
Trichlorobenzene (congener unidentified)	adipose, milk, breath	1, 8, 19
1,2,3,4-tetrachlorobenzene	adipose	32
1,2,3,5-tetrachlorobenzene	adipose	32
Pentachlorobenzene	adipose, milk	32, 25, 18
Hexachlorobenzene	adipose, milk, blood, semen	31, 16, 2,6, 18
Chloroethyl benzene	milk	9
Pentachlorophenol	semen	6
Retrachlorophenol	semen	6
Trichlorophenol	semen, urine	6, 37
3,5,6-trichloro-2-pyridinol	urine	37
Octachlorostyrene	adipose	32
4-chloro-2-(phenylmethyl)-phenoI	adipose	11
1–chloro-4-(methylsulfonyl)-benzene	adipose	11
2-chloro-6-methylbenzonitrile	adipose	11
(chlorophenyl)phenylmethanone	adipose	
Hexachloronaphthalene	semen	6, 13,7,17,14

Table 10.1 (continued)

Substance	Present in	Reference
Tetrachlorodiphenyl ether	semen	6
Hexachlorodiphenyl ether	adipose	27, 7, 17, 14
Heptachlorodiphenyl ether	adipose	27
Octachlorodiphenyl ether	adipose	27, 1, 7, 14
Nonachlorodiphenyl ether	adipose	27
Decachlorodiphenyl ether	adipose	27
Pesticides		
Aldrin	blood, adipose, milk	36, 31, 4
alpha-chlordane	adipose, milk	32, 25
Gamma-chlordane	adipose, milk	32, 25
Dieldrin	adipose, milk, blood	31, 25, 2, 37, 32, 4, 1
o,p'-DDT	adipose, blood, milk	32, 37, 25, 4
p,p'-DDT	adipose, blood, milk	1, 2, 37, 25, 32, 4
p,p'-DDE	adipose, blood, semen, milk	1, 2, 12, 37, 25, 5 32, 4
2,4-D	urine	37
Endrin	blood	35
alpha-hexachlorocyclohexane	adipose, milk, blood	18, 16, 35, 25
beta-hexachlorocyclohexane	adipose, milk, blood	1, 16, 2, 37, 25, 18
Lindane	adipose, milk	18,16,25
Heptachlor	milk	25
Heptachlor epoxide	adipose, milk, blood	31, 25, 2, 37, 32, 1, 5, 1
Mirex	adipose, blood	1, 25, 34, 32
Photomirex	milk	25
cis-nonachlor	adipose, blood	32, 37
trans-nonachlor	adipose, milk, blood	31, 16, 2, 25, 1
Nonachlor-Ill	adipose	33
Oxychlordane	adipose, milk, blood	32, 25, 2, 37
Pentachlorophenol	urine	37
Toxaphene	adipose	1
Polychlorinated dioxins (PCDDs)		
2,3,7,8-TCDD	adipose, milk, blood	1, 22, 20, 30, 37
1,2,3,7,8-PeCDD	adipose, milk, blood	1, 22, 20, 30, 37

Table 10.1 (continued)

Substance	Present in	Reference
1,2,3,4,7,8-HxCDD	adipose, milk, blood	1, 22, 20, 30, 37
1,2,3,6,7,8-HxCDD	adipose, milk, blood	1, 22, 20, 30, 37
1,2,.3,7,8,9-HxCDD	adipose, milk, blood	1, 22, 20, 30, 37
1,2,3,4,6,7,8-HpCDD	adipose, milk, blood	1, 22, 20, 30, 37
1,2,3,4,6,7,9-HpCDD	blood	37
OCDD	adipose, milk, blood	1, 22, 20, 30, 37
Polychlorinated dibenzofurans (PCDFs)		
2,3,7,8-TCDF	adipose, milk, blood	1, 22, 20, 30, 37
1,2,3,7,8-PeCDF	adipose, milk, blood	1, 22, 20, 30, 37
2,3,4,7,8-PeCDF	adipose, milk, blood	1, 22, 20, 30, 37
1,2,3,4,7,8-HxCDF	adipose, milk, blood	1, 22, 20, 30, 37
1,2,3,6,7,8-HxCDF	adipose, milk, blood	1, 22, 20, 30, 37
2,3,4,6,7,8-HxCDF	adipose, milk, blood	1, 22, 20, 30, 37
1,2,3,7,8,9-HxCDF	adipose, milk, blood	1, 22, 20, 30, 37
1,2,3,4,6,7,8-HpCDF	adipose, milk, blood	1, 22, 20, 30, 37
1,2,3)4,7,8,9-HpCDF	adipose, milk, blood	1, 22, 20, 30, 37
OCDF	adipose, milk, blood	1, 22, 20, 30, 37
Polychlorinated biphenyls (PCBs)		
2-monochlorobiphenyl	semen	12
2,2′-dichlorobiphenyl	semen	12
2,3-dichlorobiphenyl	semen	12
2,3′-dichlorobiphenyl	semen	12
2,4-dichlorobiphenyl	semen	12
2,4′-dichlorobiphenyl	semen	12
2,5-dichlorobiphenyl	semen	12
2,2′,3-trichlorobiphenyl	semen	12
2,2′,4,-trichlorobiphenyl	semen	12
2,2′,5-trichlorobiphenyl	semen	12
2,3,3′-trichlorobiphenyl	semen	12
2,3,4′-trichlorobiphenyl	semen, milk	12, 26
2,3′,4-trichlorobiphenyl	semen	12
2,3′,4′-trichlorobiphenyl	milk, blood, adipose	24, 28, 26, 32
2,4,4′-trichlorobiphenyl	adipose, milk, blood	32, 21, 28, 37, 30, 26
2,4′,6-trichlorobiphenyl	semen	12

Table 10.1 (continued)

Substance	Present in	Reference
3,4,4′-trichlorobiphenyl	milk, blood, adipose	32, 28
2,2′,3,4-tetrachlorobiphenyl	milk, adipose	24, 32
2,2′,3,5′-tetrachlorobiphenyl	semen, milk, adipose, blood	12, 24, 28
2,2′,3,6′-tetrachlorobiphenyl	milk	26
2,2′,4,5-tetrachlorobiphenyl	milk	26
2,2′,4,5′-tetrachlorobiphenyl	semen, adipose, milk, blood	12, 32, 24, 28
2,2′,5,5′-tetrachlorobiphenyl	semen, milk, blood, adipose	12, 24, 28, 32, 37
2,3,3′,4′-tetrachlorobiphenyl	milk, blood	26, 37
2,3,4,4′-tetrachlorobiphenyl	milk, adipose, blood	24, 28, 37
2,3′,4,4′-tetrachlorobiphenyl	semen, blood, milk, adipose	12, 28, 24, 32
2,3′,4′,5′-terrachlorobiphenyl	milk	26
2,3′,4′,5-tetrachlorobiphenyl	milk	26
2,4,4′,5,-tetrachlorobiphenyl	adipose, blood, milk	32, 28, 24, 37, 30, 26
3,3′,4,4′-tetrachlorobiphenyl	milk, adipose, blood	35, 37, 30
3,4,4′,5-tetrachlorobiphenyl	blood	37
2,2′,3,3′,4-pentachlorobiphenyl	semen	12
2,2′,3,4′,5-pentachlorobiphenyl	milk, blood, adipose	32, 28
2,2′,3,4,5′-pentachlorobiphenyl	semen, milk, adipose, blood	12, 26, 28
2,2′,3′,4,5-pentachlorobiphenyl	semen	12
2,2′,3,5,6′-pentachlorobiphenyl	semen	12
2,2′,4,4′,5-pentachlorobiphenyl	milk, blood, adipose	24, 28, 32, 37, 30, 26
2,2′,4,5,5′-pentachlorobiphenyl	semen, milk, adipose, blood	12, 32, 23, 37
2,3,3′,4,4′-pentachlorobuphenyl	adipose, milk, blood	32, 24, 28, 37, 30
2,3,3′,4′,5-pentachlorobiphenyt	milk	26
2,3,3′,4′,5′-pentachlorobiphenyI	milk	26, 24
2,3,3′,4′,6-pentachlorobiphenyl	milk, blood, adipose	24, 28, 28
2,3,4,4′,5-pentachlorobiphenyl	milk	26
2,3′,4,4′,5-pentachlorobiphenyl	adipose, blood, milk	32, 28, 24, 37, 30, 26

Table 10.1 (continued)

Substance	Present in	Reference
2,3′,4,4′,6-pentachlorobiphenyl	milk	26
3,3′,4,4′,5-pentachlorobiphenyl	milk, adipose, blood	35, 37, 30
2,2′,3,3′,4,4′-hexachlorobiphenyl	milk, blood	26, 30
2,2′,3,3′,4,5-hexachlorobiphenyl	milk, blood	26, 37
2,2′,3,3′,5,6′-hexachlorobiphenyl	milk	26
2,2′,3,4,4′,5′-hexachlorobiphenyl	adipose, milk, blood, semen	32, 21, 28, 12, 37, 30
2,2′,3,4,4′,5-hexachlorobiphenyl	milk	24
2,2′,3,4,5,5′-hexachlorobiphenyl	milk, adipose	24, 32
2,2′,3,4′,5,5′-hexachlorobiphenyl	milk, blood	26, 37
2,2′,3,5,5′,6-hexachlorobiphenyl	semen, milk, blood, adipose	12, 26, 28, 23
2,2′,4,4′,5,5′-hexachlorobiphenyl	semen, blood, milk, adipose	12, 28, 21, 32, 37, 30
2,3′,4,4′,5,5′-hexachlorobiphenyl	milk, blood	26, 37
2,3, 3′,4,4′,5′-hexachlorobiphenyl	adipose, milk, blood	32, 24, 28, 37, 30
2,3,3′,4,4′,5′-hexachlorobiphenyl	adipose, milk, blood	32, 28, 26, 37
2,3,3′,4,4′,6-hexachlorobiphenyl	milk	26
3,3′,4,4′,5,5′-hexachlorobiphenyl	adipose, milk, blood	35, 37, 30
2,2′,3,3′,4,4′,5-heptachlorobiphenyl	milk, adipose, semen, blood	12, 26, 29, 37, 30
2,2′,3,3′,4,4′,6-heptachlorobiphenyl	milk,semen	26, 12
2,2′,3,3′,4,5,5′-heptachlorobiphenyl	milk, blood	26, 37
2,2′,3,3′,4′,5,6-heptachlorobiphenyl	milk, blood	26, 37
2,2′,3,3′,4,5,6′-heptachlorobiphenyl	milk	26
2,2′,3,3′,4,6,6′-heptachlorobiphenyl	semen	12
2,2′,3,3′,5,6,6′-heptachlorobiphenyl	semen	12
2,2′,3,4,4′,5,5′-heptachlorobiphenyl	adipose, milk, blood	21, 28, 32, 37, 30, 24,26
2,2′,3,4,4′,5′,6-heptachlorobiphenyl	adipose, milk, blood	32, 24, 37, 30
2,2′,3,4,5,5′,6-heptachlorobiphenyl	adipose, milk, blood	26, 32, 30
2,2′,3,4′,5,5′,6-heptachlorobiphenyl	adipose, blood, milk	32, 28, 24, 37, 30
2,3,3′,4,4′,5,5′-heptachlorobiphenyl	milk, semen	12, 24, 32
2,3,3′,4,4′,5′,6-heptachlorobiphenyl	milk, blood, adipose	24, 32
2,3,3′,4′,5,5′,6-heptachlorobiphenyl	adipose, milk, blood	32, 24, 37

Table 10.1 (continued)

Substance	Present in	Reference
2,2′,3,3′,4,4′,5,5′-octachlorobiphenyl	milk, blood, adipose	24, 28, 32, 37
2,2′,3,3′,4,4′,5,6-octachlorobiphenyl	milk, blood	26, 37
2,2′,3,3′,4,4′,5′,6-octachlorobiphenyl	milk, adipose	26, 29
2,2′,3,3′,4,4′,5,6′-octachlorobiphenyl	blood, semen	37, 12
2,2′,3,3′,4,5,5′,6′-octachlorobiphenyl	semen	12
2,2′,3,3′,4′,5,5′,6-octachloobiphenyl	milk, blood, adipose	24, 28, 32
2,2′,3,3′,4,5′,6,6′-octachlorobiphenyl	blood	37
2,2′,3,3′,5,5′,6,6′-octachlorobiphenyl	milk	26
2,2′,3,4,4′,5,5′,6-octachlorobiphenyl	milk, blood, adipose	24, 28, 32, 37
2,3,3′,4,4′,5,5′,6-octachlorobiphenyl	milk	26
2,2′,3,3′,4,4′,5,5′,6- nonachlorobiphenyl	milk, blood, adipose	24, 28, 32, 37
decachlorobiphenyl	milk, blood, adipose	24, 28, 1, 37
4′-OH-chlorobiphenyl 120	blood	38
4-OH-chlorobiphenyl 109	blood	38
3-OH-chlorobiphenyl 153	blood	38
4-OH-chlorobiphenyl 146	blood	38
3′-OH-chlorobiphenyl 138	blood	38
4′-OH-chlorobiphenyl 130	blood	38
3′-OH chlorobiphenyl 187	blood	38
4-OH-chlorobiphenyl 187	blood	38
3′-OH-chlorobiphenyl 180	blood	38
4′-OH-chlorobiphenyl 172	blood	38
4-OH-chlorobiphenyl 193	blood	38

References

1. Stanley JS. *Broad scan analysis of the FY 1982 National Human Adipose Tissues Survey specimens,* Volume I: *Executive summary.* Washington DC: U.S Environmental Protection Agency, Office of Toxic Substances (EPA 560-5-86-035), 1986.

Stanley JS. *Broad scan analysis of the FY 1982 National Human Adipose Tissue Survey specimens,* Volume 3: *Semivolatile compounds.* Washington DC: U.S. Environmental Protection Agency (EPA 560/5–86/037), 1986.

2. Murphy R, Harvey C. Residues and metabolites of selected persistent halogenated hydrocarbons in blood specimens from a general population survey. *Environmental Health Perspectives* 60:115–120, 1985.

3. Kutz FW, Strassman SC, Sperling JF. Survey of selected organochlorine pesticides in the general population of the United States: Fiscal years 1970–1975. *Annals of the New York Academy of Science* 31;320:60–68, 1979.

4. Jensen AA. Chemical contaminants in human milk. *Residue Reviews* 89:1–128, 1983.

5. Savage EP, Keefe TJ, Tessari JD, Wheeler HW, Applehans FM, Goes EA, Ford SA. National study of chlorinated hydrocarbon insecticide residues in human milk, USA: I. Geographic distribution of dieldrin, heptachlor, heptachlor epoxide, chlordane, oxychlordane, and mirex. *American Journal of Epidemiology* 113(4):413–422, 1981.

6. Dougherty RC, Whitaker MJ, Tang SY, Bottcher R, Keller M, Kuehl DW. Sperm density and toxic substances: A potential key to environmental health hazards. In: McKinney JD, ed. *The chemistry of environmental agents as human hazards.* Ann Arbor, MI: Ann Arbor Science Publishers 1981:263–277.

7. Wallace LA, Pellizzari E, Hartwell T, Rosenzweig M, Erickson M, Sparacino C, Zelon H. Personal exposure to volatile organic compounds. I. Direct measurements in breathing-zone air, drinking water, food, and exhaled breath. *Environmental Research* 35:293–319, 1984.

8. Pellizzari ED, Hartwell TD, Harris BS, Waddell RD, Whitaker DA, Erickson MD. Purgeable organic compounds in mother's milk. *Bulletin of Environmental Contamination and Toxicology* 28:322–328, 1982.

9. Jensen AA. Polychlorobiphenyls (PCBs), polychlorodibenzo-p-dioxins (PCDDs) and polychlorodibenzofurans (PCDFs) in human milk, blood and adipose tissue. *Science of the Total Environment* 64(3):259–293, 1987.

10. Hudec T, Thean J, Kuehl D, Dougherty RC. Tris(dichloropropyl)phosphate, a mutagenic flame retardant: Frequent occurrence in human seminal plasma. *Science* 211:951–952, 1981.

11. Onstot J, Ayling R, Stanley J. *Characterization of HRGC/MS unidentified peaks from the analysis of human adipose tissue,* Volume I: *Technical approach.* Washington DC: U.S. Environmental Protection Agency, Office of Toxic Substances (560/6-87-002a), 1987.

12. Bush B, Bennett AH, Snow JT. Polychlorinated biphenyl congeners, p,p′-DDE, and sperm function in humans. *Archives of Environmental Contamination and Toxicology* 15:333–341, 1986.

13. Laseter JL, Dowty BJ. Association of biorefractories in drinking water and body burden in people. *Annals of the New York Academy of Sciences* 298:547–556, 1978.

14. Wallace LA, Pellizzari ED, Hartwell TD, Whitmore R, Zelon H, Perritt R, Sheldon L. The California TEAM study: Breath concentrations and personal exposures to 26 volatile compounds in air and drinking water of 188 residents of Los Angeles, Antioch, and Pittsburg, CA. *Atmospheric Environment* 22:2141–2163, 1988.

15. Onstot J, Stanley J. *Identification of SARA compounds in adipose tissues.* Washington DC: U.S. Environmental Protection Agency, Office of Toxic Substances (EPA-560/5-89-003), 1989.

16. Takei GH, Kauahikaua SM, Leong GH. Analyses of human milk samples collected in Hawaii for residues of organochlorine pesticides and polychlorobiphenyls. *Bulletin of Environmental Contamination and Toxicology* 30:606–613, 1983.

17. Krotoszynski BK, Bruneau GM, O'Neill JH. Measurement of chemical inhalation exposure in urban population in the presence of endogenous effluents. *Journal of Analytical Toxicology* 3:225–234, 1979.

18. Mes J, Davies DJ, Turton D. Polychlorinated biphenyl and other chlorinated hydrocarbon residues in adipose tissue of Canadians. *Bulletin of Environmental Contamination and Toxicology* 28:97–104, 1982.

19. Krotoszynski BK, O'Neill JH. Involuntary bioaccumulation of environmental pollutants in nonsmoking heterogeneous human population. *Journal of Environmental Science and Health* A 17:855–883, 1982.

20. Schecter A, Gasiewicz T. Health hazard assessment of chlorinated dioxins and dibenzofurans contained in human milk. *Chemosphere* 16:2147–2154, 1987.

21. Schecter A, Furst P, Kruger C, Meemken HA, Groebel W, Constable JD. Levels of polychlorinated dibenzofurans, dibenzodioxins, PCBs, DDT, DDE, hexachlorobenzene, dieldrin, hexachlorocyclohexanes and oxychlordane in human breast milk from the United States, Thailand, Vietnam, and Germany. *Chemosphere* 178:445–454, 1989.

22. Kahn PC, Gochfeld M, Nygren M, Hansson M, Rappe C, Velez H, Ghent-Guenther T, Wilson WP. Dioxins and dibenzofurans in blood and adipose tissue of Agent Orange–exposed Vietnam veterans and matched controls. *Journal of the American Medical Association* 259:1661–1667, 1988.

23. Mes J, Marchand L. Comparison of some specific polychlorinated biphenyl isomers in human and monkey milk. *Bulletin of Environmental Contamination and Toxicology* 39:736–742, 1987.

24. Mes J, Marchand L. Comparison of some specific polychlorinated biphenyl isomers in human and monkey milk. *Bulletin of Environmental Contamination and Toxicology* 39:736–742, 1987.

25. Davies D, Mes J. Comparison of the residue levels of some organochlorine compounds in breast milk of the general and indigenous Canadian populations. *Bulletin of Environmental Contamination and Toxicology* 39:743–749, 1987.

26. Safe S, Safe L, Mullin M. Polychlorinated biphenyls: Congener-specific analysis of a commercial mixture and a human milk extract. *Journal of Agricultural and Food Chemistry* 33:24–29, 1985.

27. Stanley JS, Cramer PH, Ayling RE, Thornburg KR, Remmers JC, Breen JJ, Schwemberger J. Determination of the prevalence of polychlorinated diphenyl ethers (PCDPEs) in human adipose tissue samples. *Chemosphere* 20:981–985, 1990.

28. Mes J, Weber D. Non-orthochlorine substituted coplanar polychlorinated biphenyl congeners in Canadian adipose tissue, breast milk and fatty foods. *Chemosphere* 19:1357–1365, 1989. Schecter A et al. Polychlorinated biphenyl (PCB), DDT, DDE, and hexachlorobenzene (HCB) and PCDD/F isomer levels in various organs and autopsy tissue from North American patients. *Chemosphere* 18:811–818, 1989.

29. Williams DT, LeBel GL. Polychlorinated biphenyl congener residues in human adipose tissue samples from five Ontario municipalities. *Chemosphere* 20:33–42, 1990.

30. Schecter A, Papke O, Ball M. Evidence for transplacental transfer of dioxins from

mother to fetus: Chlorinated dioxin and dibenzofurans levels in the livers of stillborn infants. *Chemosphere* 21:1017–1022, 1990.

31. Kutz FW, Wood PH, Bottimore DP. Organochlorine pesticides and polychlorinated biphenyls in human adipose tissue. *Reviews of Environmental Contamination and Toxicology* 120:1–82, 1991.

32. Mes J, Marchand L, Davies DJ. Organochlorine residues in adipose tissue of Canadians. *Bulletin of Environmental Contamination and Toxicology* 45:681–688, 1990.

33. Dearth M, Hites RA. Highly chlorinated dimethanofluorenes in technical chlordane and in human adipose tissue. *Journal of the American Society for Mass Spectrometry* 1:92–98, 1990.

34. Moysich KB, Ambrosone CB, Vena JE, Shields PG, Mendola P, Kostnyiak P, Grezerstein H, Graham S, Marshall JR, Schisterman EF, Freudenheim JL. Environmental organochlorine exposure and postmenopausal breast cancer risk. *Cancer Epidemiology Biomarkers and Prevention* 7:181–188, 1998.

35. Djordevic MV, Hoffman D, Fan J, Prokopczyk B, Citron ML, Stellman SD. Assessment of chlorinated pesticides and polychlorinated biphenyls in adipose breast tissue using a supercritical fluid extraction method. *Carcinogenesis* 15:2581–2585, 1994.

36. Mossing ML, Redetzke KA, Applegate HG. Organochlorine pesticides in blood of persons from El Paso, Texas. *Journal of Environmental Health* 47:312–313.

37. Anderson HA, Falk C, Hanrahan L, Olson J, Burse VW, Needham L, Paschal D, Patterson D, Hill RH, Great Lakes Consortium. Profiles of Great Lakes critical pollutants: A sentinel analysis of human blood and urine. *Environmental Health Perspectives* 106:279–289, 1998.

38. Sandau CD, Ayotte P, Dewailly E, Duffe J, Norstrom RJ. Analysis of hydroxylated metabolites of PCBs (OH-PCBs) and other chlorinated phenolic compounds in whole blood from Canadian Inuit. *Environmental Health Perspectives* 108:611–616, 2000.

primarily targeted substances that are already the focus of scientific and regulatory attention, such as PCBs, dioxins, and DDT and other pesticides.

Three biomonitoring surveys have studied broad samples of the U.S. population. First, the National Health and Nutrition Examination Survey (NHANES II), lead and pesticide residues in blood were studied during the period 1976 to 1980. In this national sample of nearly 6,000 people aged 12 to 74 years, 36 pesticides and pesticide metabolites were quantified.[25,26] Second, in 1981, Savage and colleagues reported pesticide levels in human breast milk in a large sample of the U.S. nursing-mother population.[27] Third, using surgical and postmortem body-fat specimens, the Environmental Protection Agency's National Human Adipose Tissue

Survey (NHATS) estimated the general population's body burden of a variety of synthetic chemicals.[28–30] From the early 1970s to 1992, the survey collected and analyzed a national sample of adipose-tissue specimens for a variety of organic substances.[29] Initially, the NHATS chemical analysis included PCBs and organochlorine pesticides; in 1982, the chemical analysis was expanded to include several dozen additional volatile and semivolatile Toxic Substances Control Act–related chemicals.[28] An additional broad-scan research effort sought to characterize adipose-contained pollutants that were not on the target list; 298 additional contaminants were identified, but 601 semivolatile and 99 volatile substances could not be characterized.[15] The NHATS programs has been criticized for lacking a standardized methodology and using a sample of individuals that may not accurately reflect the nation's population, but the program's results remain one of the most comprehensive available data sets on the general population's body burden.[12]

Two surveys have sought to identify a large number of exogenous substances in the general human population, but the study groups were small and may not have been fully representative of the nation at large. EPA's TEAM study sought to measure a large number of substances in human breath in 12 volunteers from New Jersey and North Carolina.[31] Another EPA analysis identified a large number of synthetic chemicals in mother's milk from 42 women in New Jersey, Pennsylvania, and Louisiana.[32]

In 1991, a committee of the National Academy of Sciences called for a national program of monitoring body fluids from the general population for the presence of a list of target substances, based on a standardized protocol.[12] Subsequently, the National Center for Environmental Health (NCEH), a division of the U.S. Centers for Disease Control and Prevention (CDC), initiated the National Report on Human Exposure to Environmental Chemicals. Each year, the NCEH biomonitoring laboratory will measure and report the exposure of a representative sample (n = 1,200) of the U.S. population—a subsample of the ongoing National Health and Nutrition Examination Survey (NHANES)—to a number of priority toxic substances. Adults and children as young as six will be studied. In 2001, the CDC issued its first report from the National Report on Human Exposure to Environmental Chemicals program. This report contains information about the exposure of the U.S. population to 27 environmental chemicals, including 13 heavy metals, cotinine (a metabolite of nicotine), 6 metabolites of organophosphate pesticides (from the breakdown of 28 pesticides), and 7 phthalate metabolites. By comparing the results with future reports, public health officials will be able to determine how health is being affected as levels of exposure to environmental chemicals change. Data are reported according to the age, sex, race/ethnicity, geographic area, and income level. Data are publicly available and will include frequency distributions of contaminant levels and demographic data.[33]

In the second year, the report card will include an additional 25 compounds, including volatile organic compounds, gasoline additives, and major dioxin, furan, and PCB congeners. In the third year, the number of measured compounds will increase to 100. Access to the biomonitoring results of the National Exposure Report Card will enhance the ability of the American public to learn about its body burden of industrial chemicals and the ability of scientists to study health risks of chemical exposures.

A particular value of this national biomonitoring program is its capacity to a establish reference ranges for a large set of toxic chemicals. A reference range is defined as the concentration of a particular substance that is expected to be present in the general population with no unusual chemical exposure.[34] The reference range is the standard against which a measuring laboratory can say that results for any group or individual are high, in a "normal" range, or low.[35] For materials like lead, cadmium, and some pesticides, good reference values exist because large numbers of persons have been studied using well-standardized and reproducible methods. For many other substances, laboratory methods have varied over time, and few large nationally representative populations have been studied. Reference-range data for heavy metals,[36] dioxins and PCBs,[37] and selected pesticides[34,38] have been published.

Organochlorine Substances

Organochlorines are a class of carbon-based (organic) chemicals that contain one or more chlorine atoms. About 40 million tons of chlorine are produced annually by the world chlor-alkali industry.[39] Over three-quarters of this amount is used in the production of more than 11,000 organochlorine chemicals, including plastics, pesticides, solvents, and chemical intermediates; thousands of additional organochlorines are formed as by-products in the manufacture, some uses, and disposal of organochlorine-containing products.[4] Organochlorines are not the only industrial chemicals of concern; organobromines, heavy metals, organophosphates, and aromatic compounds (including benzene, phthalate esters, bisphenol, alkyl phenols, and polynuclear aromatic hydrocarbons) are also widespread pollutant classes that pose public health risks when present in human tissues. But organochlorines dominate all lists of priority hazardous pollutants, including the list of 12 persistent organic pollutants (POPs) slated for action under the first international agreement on POPs[40] and the lists of pollutants of concern in the Great Lakes ecosystem.[41] The American Public Health Association has recommended that organochlorines be treated as a class and phased out wherever safe alternatives exist.[42]

Three characteristics of organochlorines make them particularly troublesome. First, chlorination changes the chemical stability of organic chemicals in largely predictable ways, making many organochlorines highly persistent in the environment.[4] Even dilute discharges of such substances accumulate in the environment over time, reaching measurable levels in air, water, and sediments. Persistent substances are transported long distances through the atmosphere and are distributed globally, accumulating even in remote regions such as the Canadian and European Arctic, the rain forests of South America, and the Pacific islands.[6,22,43]

Second, many organochlorines are strongly lipophilic—that is, they are highly soluble in fats but not in water, so they accumulate in fatty tissues. Chlorination almost invariably increases the solubility of organic substances in fat and reduces their water solubility; the increase in lipophilia becomes greater with each chlorine atom added to the molecule.[4,44] In addition, fat-soluble substances that resist degradation and excretion magnify in concentration as they move up the food chain; organochlorine body burdens in carnivores are typically millions of times greater than the levels found in ambient air, soil, and sediments.[45] The human population, at the apex of the food chain, is particularly contaminated.[4,22]

Finally, organochlorines tend to be considerably more toxic and more carcinogenic than their nonchlorinated analogues.[10] It has long been known that some organochlorine chemicals produce a variety of adverse health impacts, including cancer, organ damage, and reproductive, neurological, and immunological toxicity, particularly during sensitive periods of fetal and infant development.[4,10,42] Several hundred organochlorines have undergone toxicological testing,[2] and virtually all organochlorines examined to date have been found to cause one or more of a wide variety of adverse effects, often at very low doses.[35,46,47] A large number of organochlorines are endocrine disrupters that mimic or otherwise interfere with hormone action, raising the possibility of severe long-term effects on reproduction, development, and behavior; [48] the National Academy of Sciences has recently concluded that hormonally active agents (HHAs) in the environment can affect human health, but the extent to which public health has been harmed is controversial.[9]

"Normal" Body Burdens

All humans are now exposed to organochlorines and other industrial chemicals in drinking water, air, and food. For bioaccumulative compounds, including dioxins, PCBs, and many pesticides, the food supply is the major source of exposure. Over 90 percent of the average American's dioxin exposures, for example, come through the diet. Animal foods—meat, dairy, fish, and eggs—contain the highest concentrations of dioxin, but smaller exposures also occur through grains, fruits, and vegetables and their oils.[49] For the more volatile and less bioaccumulative chemicals, such as solvents, disinfection by-products, and pesticides in groundwater, exposures via air and water play more important roles.[50–52] Direct exposures to organochlorines also occur via consumer products, home pesticides, and dry-cleaned clothing.[53–55]

As a result of these universal exposures, the general population now carries a body burden of numerous organochlorines. Over 190 synthetic organochlorines have been identified in samples of tissues and fluids drawn from the general populations of the United States and Canada (see table 10.1). The studies on which this list is based are subject to the limitations discussed above, but they do indicate that a very large number of synthetic substances found in the environment and food supply have accumulated in the bodies of individuals who have had no special chemical exposures. As one might expect, the list includes the best-studied organochlorines: dioxins, PCBs, and restricted pesticides. Like the ambient environment, however, the bodies of the general human population contain a representative sample of the full range of organochlorines, from simple chlorinated refrigerants and solvents used for cleaning, coating, and degreasing, to little-known specialty chemicals, by-products, and second-generation pesticides. The concentrations are typically small, ranging from the low parts per trillion to the high parts per billion, but they indicate widespread exposure and a continuing internal dose of a variety of compounds, many of which are known to cause health damage.

The list in table 10.1 is only a subset of the pollutants that have accumulated in the tissues and fluids of the general population. The table excludes studies conducted on populations other than North Americans, and numerous pollutants have been discovered in studies from other nations. For example, a recent study found parts-per-billion concentrations of the by-products tris(4-chlorophenyl)methane and tris(4-chlorophenyl)methanol fire retardants in samples of adipose tissue drawn from individuals in the general population of Japan.[56] Moreover, it is virtually certain that a large number of additional organochlorines—likely in the hundreds or thousands—are present but have not yet been identified due to the limitations of analytic protocols, as discussed above. Finally, there are a large number of nonchlorinated substances that contaminate the human population that are not included in table 10.1. For example, heavy metals accumulate in bone and are detectable in blood and urine.[36,57] Hydrocarbon solvents are commonly found in fat and blood, with toluene and xylene detected in 100 percent of NHANES III subjects.[58] Pesticide residues including nonorganochlorines have been found in fat, blood, and urine.[34,37,38,59] Parts-per-million concentrations of metabolites of phthalate esters, reproductive toxicants that are used in vinyl plastics and other products, have recently been identified in the urine of the general U.S. population.[60]

It it now "normal" for the general population to have body burdens of industrial chemicals and metals, in the sense that no reference or control population can be found without these exposures. But the existence of an appreciable body burden of industrial chemicals is a relatively recent phenomenon. Lead and other metals have been measured in human tissue since the 1920s.[57] Dioxin is nondetectable or present at only a very small fraction of current concentrations in preserved human tissue samples from preindustrial times.[61–63]

Body-Burden Variation

There is very limited published information about how body burdens of industrial chemicals vary with age, gender, geography, race, occupation, and time.

Age

Tissue concentrations of industrial chemicals that bioaccumulate are expected to continually increase with age, a prediction that has been corroborated by numerous studies. Orban and colleagues reported from 1987 NHATS data that the average body-burden levels of PCDDs and PCDFs in the general population increase with age, with the highest levels being seen in the 45-plus year-old group; concentrations increased by 1 to 2 pg/g per decade.[28] Other studies of small and less representative populations have shown similar results.[26,28,59,64–67] Pollutant levels have been studied in single subjects at several points in time to determine the biological half-life of specific chemicals.[68] However, no long-term, longitudinal studies have been conducted to track the accumulation of persistent chemicals with the passage of time in an individual or group.

Gender

Sex differences in pollutant exposure and pharmacokinetics that may affect pollutant accumulation have not been well studied. One important determinant of body-burden concentrations that has been investigated to some extent is the excretion of bioaccumulated pollutants in mother's milk.[69] In nonhuman mammals, females have body burdens of highly persistent organochlorines that are only half as high as those of males because they transfer to their young as much as 60 percent of the pollutants they have accumulated in their own tissues.[70] In the NHANES II study of 5,000 persons in 1981, men of all ages had higher residues of 16 pesticides than women.[26] But in the 1982 and 1987 NHATS, there was no statistically significant difference between the genders in body burdens of dioxins and furans in the general population.[28]

Geography

Although industrial chemicals are ubiquitous, location affects the magnitude of exposure. In North America, atmospheric currents and global distillation affect transport of semivolatile substances like PCBs and pesticides from the U.S. North to the Canadian arctic, where they accumulate in surprisingly high concentrations.[6,71] As a result of this transport and a diet in rich in the blubber of marine mammals, the body burdens of PCBs and numerous organochlorine pesticides in milk from Inuits in arctic Quebec are two- to tenfold higher than those of urban residents of southern Canada.[72]

There are also regional differences within the United States in body burdens of some chemicals, with higher body burdens in regions with greater environmental contamination. In the NHATS program, adipose-tissue levels of pesticides and PCBs were highest in regions where levels in fish and sediments were high; [73] there were no significant differences in PCDD body burdens among the four U.S. census regions.[28,74] Body burdens of PCDDs appear to be higher in studies conducted in New Jersey and Missouri than in California, but the results may be due to measurement or population differences.[20,26,75]

Finally, local pollutant sources can cause increased exposures. Community surveys of pesticide exposure suggest that residents of rural agricultural communities have higher body burdens than urban residents.[26] Individuals living near a chemical plant in Louisiana that produces organochlorine feedstocks for polyvinyl chloride plastics have significantly elevated levels of dioxins and furans in their blood.[76]

Race

There are notable racial differences in the body burden of organochlorine pesticides.[59] In 1968, U.S. EPA data from 22 U.S. states showed significantly higher levels of DDE in adipose tissues of African Americans than in whites.[77] Subsequently, Kutz and colleagues found that median DDE in human adipose tissues from African Americans was twice that of whites in general, nonoccupationally exposed populations.[78] In the 1981 NHANES survey, nonwhites were more likely to have detectable residues of 16 pesticides tested and had higher DDE levels.[26] Blood-lead levels have been consistently found to be higher in African Americans than in whites.[79,80] For most substances, however, there are no data on racial differences in body burdens.[81]

Occupation

There are ample data to confirm that persons with occupational exposures have higher body burdens of industrial chemicals than persons without such occupational exposures, chemical and agricultural workers being the obvious examples.[1,16,26,35,82]

Time Trends

Efforts to track trends in body burdens are undermined by a lack of historical data for most pollutants and, for substances for which data from previous decades are available, changes in analytic methods that make results noncomparable. Nevertheless, many studies confirm that concentrations of organochlorine chemicals that were restricted in the United States and Canada—such as PCBs, DDT, dieldrin, heptachlor, and several other pesticides—declined in adipose tissue and mother's milk in the 1970s and 1980s and then stabilized.[34,59,69,83] Over the four-year period of NHATS (1982–1986), trends in organochlorine pesticide residues (e.g., DDT, DDE, nonachlor, and heptachlor) were quite stable despite the fact these agents were restricted in the 1970s, presumably because of the extraordinary environmental persistence of these substances.[34] Dioxin levels in human breast milk also declined in the 1970s and 1980s; levels fell in many European nations during the 1990s as well, although the data in the United States are ambiguous.[69,84] The many other organochlorines that were not subject to severe restrictions presumably did not follow the same pattern. Detection of the insecticide chloropyrifos in the urine of the general U.S. population, for example, increased from 7 percent in 1980 to 52 percent in 1990.[85] For most compounds, there are no data on time trends in human-tissue concentrations.

Critical Populations: Women and Children

Once persistent organochlorines have taken up residence in human fatty tissues, there is no commonly recognized therapeutic way to get rid of them. Women, however, excrete accumulated persistent chemicals into the fat of breast milk and, in smaller quantities, into the fetus transplacentally.[86–88] Fifteen pesticides, 17 dioxins and furans, 64 PCBs, 11 other aromatic organochlorines, and 18 solvents and other chlorinated aliphatic substances have been identified in the breast milk of the general U.S. and Canadian populations (table 10.1). The nursing infant is exposed to these compounds during critical periods of development, when sensitivity to chemical disruption is particularly high.[9,89]

Babies' chemical exposures begin in utero. One function of the placenta is to serve as a barrier to keep harmful substances from the embryo and fetus, but fat-soluble organochlorines cross the placenta readily. Dioxins and PCBs, for example, have been found in samples of fetal organs, placentas, umbilical-cord blood, and amniotic fluid from the general U.S. population.[90–93] As a result, the fetus is exposed to biologically active organochlorines during exquisitely sensitive periods of development.[89]

A child's exposure to synthetic chemicals accelerates after birth. Just as pollutant concentrations multiply with each step up the food chain, a breast-fed baby is subject to doses of bioaccumulative organochlorines that are considerably higher than those the mother received. An average nursing infant in the United States receives a daily dose of dioxins and PCBs that is 92 times higher than that of an average adult.[67] In a year of nursing, a typical baby receives 12 percent of his or her entire lifetime dioxin exposure and accumulates a higher body burden than an average 70-year-old man.[49] The nursing child's dioxin dose exceeds by several-fold the standard of most governments in the world for an adult's "acceptable daily intake."[64,90,94,95] In the Arctic, where levels of bioaccumulative organochlorines in human milk are several-fold higher than at temperate latitudes, the problem is compounded by the fact that Inuit women nurse their children for a relatively long period.[22] A breast-fed Inuit infant's daily doses of dioxins, PCBs, toxaphene, and hexachlorobenzene are 2 to 48 times the Canadian government's "tolerable daily intake" for adults.[72]

After weaning, children continue to receive unusually high exposures, due to higher intake rates of food as a fraction of body weight. The average daily dose of dioxin-like compounds for children under age 5 in the U.S. population is four times greater than that of the average adult.[67] There have been no comprehensive analyses of the chemical body burdens of children, although some tissue samples from cadavers and surgical patents ages 0 to 14 years were analyzed in the NHATS program.[28] Given the sensitivity of developing endocrine, neurobehavioral, and reproductive systems to chemical insult, this lack of data is a matter of considerable concern.[89] A large national prospective longitudinal cohort study of American children has recently been proposed.[96] This "Children's Framingham" study would investigate the role of various risk factors, including early exposure to environmental toxicants, in impaired neurobehavioral development. A large study of pollutants in the milk of U.S. mothers has also been proposed.[69]

Public Health and Epidemiologic Significance

The measurement of body burden provides an assessment of an individual's long-term exposure to a chemical, a critical parameter in epidemiology and risk assessment. For bioaccumulative substances, body-burden measurements integrate exposures that occur across time and environmental media, and they reflect the accumulation of pollutants after metabolic and partitioning processes have taken place. Body-burden data can thus play a critical role in identifying novel hazards and high-risk populations, tracking trends in human exposure, and characterizing exposure levels that pose health hazards.[12] Body-burden measurements automatically account for differences between species in metabolism and excretion, increasing confidence in interspecies and interindividual estimates of toxicity; it is for this reason that EPA chose to make body burden the relevant dose metric in its dioxin risk assessment.[67]

The xenobiotic body burden is biologically important because it provides a reservoir for continuous internal exposure to these substances. Bioaccumulated substances exist in a dynamic equilibrium between stable biological matrices, such as fat and bone, and circulating materials, such as blood. Long-term internal exposure due to the accumulation of past exposures in an individual's tissues has been shown to have the potential to produce chronic health impacts, including tumor promotion.[97] The passage of stored pollutants to future generations makes the chemical body burden critical from a public health perspective as well.

Measurement of body burdens also provides an important means to study the relation between universal chemical exposure and adverse health outcomes.[11] Citing the absence of conclusive epidemiologic evidence, some commentators have concluded that "background" levels of chemical pollution and exposure are not causing health damage in the general population.[98,99] With no uncontaminated control group anywhere on the planet, however, traditional epidemiologic approaches are poorly suited to providing evidence that meets this standard, for several reasons. First, when the study population is the general public, there is no less-exposed group to serve as a comparison population, so it is structurally impossible to directly demonstrate health impacts at "background" exposure levels.[100] It is therefore necessary to compare groups from the general population that differ slightly or moderately in the magnitude of their exposures. With small differences between study and reference groups, accurate estimates of exposure are critical but have proven extremely troublesome. Exposures change over time, include hundreds or thousands of individual substances, and may occur 30 years or more before health impacts become manifest, sometimes in the offspring of the exposed individual. Finally, effects must also be measured accurately, but relevant impacts (e.g., deficits in immunity, fertility, or cognition) are often subtle, difficult to quantify, vary naturally within the population, and reflect exposure to chemical mixtures, nonchemical agents, and other confounders. The analytic tools now available to epidemiology are therefore not capable of completely untangling the causal webs that link long-past complex exposures to subtle forms of large-scale health damage. It is unrealistic to expect conclusive evidence of specific causal links between universal exposures and health impacts in the general population, though such effects may be of considerable public health significance.

Studies based on body burden have the potential to help resolve some of these problems. First, body burdens provide a present indicator of long-term exposure, and analyses can target an individual substance or pursue a broad scan of many pollutants. Second, body burdens in the human population can be compared to those that produce adverse health impacts in more easily studied laboratory animals. Finally, statistical relationships can be evaluated between variations in the body burden of specific pollutants or mixtures in the general public and the risk of specific health endpoints.

Health Effects at Background Exposures

For some substances, inferences based on body-burden data provide a sound basis for concern that public health may be adversely affected at "background" exposure levels, although the data remain, for the reasons discussed above, less than conclusive. First, current "normal" body burdens of dioxin and several other well-studied organochlorines in humans are at or near the range at which toxic effects occur in laboratory animals. The body burden of dioxins and furans in the average American has been estimated at 8 to 13 parts per trillion, expressed as TCDD-equivalents (TEQ);[7,28] when dioxinlike PCBs are included, the total TEQ in blood lipids in the 1990s is 25 parts per trillion.[67] In comparison, a dioxin dose that produces a body burden of just 5 parts per trillion in blood lipids of pregnant rats reduces sperm density in the male offspring by 25 percent. At a maternal body burden of 13 parts per trillion, puberty is delayed and penises and ducts in the testes are smaller. At 56 to 64 parts per trillion—higher than average but still in the range of some individuals in the general population—sexual behavior of the offspring is demasculinized, sperm density is lower by 50 percent, and prostates, testicular structures, and pituitary glands are all smaller than normal.[47] PCB levels in the general population are about the same as those that produce permanent transgenerational immunosuppression in rhesus monkeys[101] and are predicted to cause mild hypothyroidism during development based on extrapolation from the levels that cause the same effect in rodents.[102] Blood levels of the pesticides o,p′-DDT and beta-hexachlorocyclohexane in the general population are in the range that produces classic estrogenic responses in female rodents.[103]

Limited epidemiologic evidence from the general human population also suggests that background exposure to some industrial chemicals may be causing adverse health impacts.[7,102,104,105] Consider, for example, studies of the impact of prenatal and lactational exposure to organochlorines on neurodevelopment. A long-term study in Michigan has compared the developmental status of children born to mothers in Michigan who ate a moderate amount of fish—two to three meals per month—from the Great Lakes to that of a reference group of children from the same communities who did not eat Great Lakes fish. The children of fish-eating mothers had impaired performance in visual recognition tasks at five and seven months of age and reduced performance in short-term memory on both verbal and quantitative tests through 11 years of age. Most important, the severity of the cognitive effects correlated with the concentrations of PCBs in the umbilical cord serum of the children at birth.[92,106,107] Similarly, a series of studies of women from the general population in North Carolina found that higher levels of PCBs and DDE in mother's milk were associated with poor muscle tone and altered reflexes in their infants at birth. At ages six months and one year, children with higher contaminant levels scored poorly on psychomotor tests that measure muscle control and visual recognition memory. The effect persisted at ages 18 and 24 months, but it was no longer significant in tests at 5 and 10 years of age.[108–111]

Finally, an ongoing study of newborns from the general population of the Netherlands found that prenatal PCB exposure, determined as the burden of total PCBs in the mothers' plasma, was related to lower birth weight, lower growth rate, lower psychomotor scores at 3 months of age, and reduced performance on neurological tests at birth and 18 months. Maternal dioxin and furan levels were correlated with lower levels of thyroid hormones in the infants' blood at ages 1 week and 3 months, providing a plausible mechanism for the neurological effect. At 42 months, performance on several cognitive tests was inversely related to total PCB levels in the mother's plasma, indicating that "in utero exposure to background PCB concentrations is associated with poorer cognitive functioning in preschool children."[112–117]

Other studies have used similar logic to suggest links between the organochlorine body burden and other health impacts. Maternal PCB levels are strongly correlated with the incidence of low birth weight among children born to women who eat fish from the Baltic Sea.[118–120] The risk of newborn hemorrhagic disease has been linked to dioxin levels in the breast milk of mothers from the general Dutch population.[121,122] A number of studies have found significant associations between the risk of breast[123–130] and hematopoietic cancers[131–134] and the body burden of PCBs or specific organochlorine pesticides, although other studies have not corroborated these results.[135–137] Finally, maternal body burdens of PCBs and dioxin are associated with reduced thyroid hormone levels in Dutch infants.[138–140]

These findings underscore the potentially profound public health issues raised by universal human exposure to complex mixtures of anthropogenic contaminants. Every human individual on the planet is exposed, from the moment of conception until death, to a diverse mixture of industrial chemicals that can disrupt critical biological functions. Health damage may occur in individuals as the result of exposure during critical periods of development and/or may emerge slowly as a shift in the distribution of functional capacity and characteristics in the entire population. Given the limits of epidemiologic tools, the incomplete characterization of exposures and effects, and the severity and immense scale of the potential impacts, we believe it is appropriate for suggestive evidence of the sort reviewed above to receive serious consideration in both risk assessment and policymaking.

Public Understanding of Body-Burden Data

Body-burden measurements can play an important role in risk communication and provide workers and citizens with knowledge of their personal exposures. When persons understand the sources and pathways of exposure, they may modify their diet or change their residence or occupation. They may ask for health screening and traditional or alternative medical advice about their risk factors. They may take political action and seek systematic reductions in the release of pollutants to the environment.

There are no public opinion-survey data that indicate what portion of the general population of Americans are aware of the chemical body burden, or whether, if people know of their personal exposure, they are concerned. We do know that some groups—workers, cancer survivors, environmentalists, and communities near toxic industrial facilities and incinerators—are aware of these issues and are active in efforts to reduce contamination and exposure.

Health knowledge by itself can change behaviors. Smokers or persons who practice unprotected sex are more likely to change their behaviors if they have knowledge of the risks of heart disease or HIV infection than others without such information. In many cases the effects of health education are modest, but health-promotion programs are based on evidence that health knowledge is positively associated with appropriate health behaviors. Public knowledge of blood-lead levels has been important in efforts to promote lead screening and reduce exposure. Some chemical workers and farmers have changed their practices after learning their blood levels of pesticides and industrial chemicals. Information on the universal chemical body burden may be an important tool in pollution prevention and environmental health education for the general public.

References

1. Rom WN. *Environmental and occupational medicine*. 2nd ed. Boston: Little, Brown, 1992.

2. Roe D, Pease W, Florini K, Silbergeld E, Leiserson K, Below C, et al. *Toxic ignorance: The continuing absence of basic health testing for top-selling chemicals in the United States*. NewYork: Environmental Defense Fund, 1997.

3. Yang RSH. *Toxicology of chemical mixtures*. New York: Academic Press, 1994.

4. Thornton J. *Pandora's poison*. Cambridge, MA: MIT Press, 2000.

5. Loganathan BG, Kannan K. Global organochlorine contamination trends: An overview. *Ambio* 1994; 23:187–191.

6. Simonich SL, Hites RA. Global distribution of persistent organochlorine compounds. *Science* 1995; 269:1851–1854.

7. DeVito MJ, Birnbaum LS, Farland WH, Gasiewicz TA. Comparisons of estimated human body burdens of dioxinlike chemicals and TCDD body burdens in experimentally exposed animals. *Environ Health Perspect* 1995; 103:820–831.

8. DeVito MG, Birnbaum LS. Toxicology of dioxins and related chemicals. In: Schecter A, editor. *Dioxins and health*. New York: Plenum, 1994. pp. 139–162.

9. Committee on Hormonally Active Agents in the Environment, Board on Environmental Studies and Toxicology, Commission on Life Sciences, National Research Council. *Hormonally active agents in the environment*. Washington, DC: National Academy Press, 1999.

10. Henschler D. Toxicity of chlorinated organic compounds: Effects of the introduction of chlorine in organic molecules. *Angewandte Chemie International Edition* (English) 1994; 33:1920–1935.

11. Landrigan PJ. Relation of body burden measures to ambient measures. In: Gordish L, Libauer CH, editors. *Epidemiology and health: Risk assessment*. New York: Oxford University Press, 1988. pp. 139–147.

12. National Research Council, Committee on National Monitoring of Human Tissues. *Monitoring human tissues for toxic substances*. Washington, DC: National Academy Press, 1991.

13. Agency for Toxic Substances and Disease Registry. *Toxicological profile series*. Washington, DC: US Public Health Service, 1989–1999.

14. Committee on Biological Markers of the National Research Council. Biological markers in environmental health research. *Environ Health Perspect* 1987; 74:3–9.

15. Onstot J, Ayling R, Stanley J. *Characterization of HRGC/MS unidentified peaks from the analysis of human adipose tissue*. Vol 1: *Technical approach*. Washington, DC: US Environmental Protection Agency, Office of Toxic Substances, 1987. 560/6–87–002a.

16. Wolff MS, Anderson HA. Polybrominated biphenyls: Sources and disposition of exposure among Michigan farm residents, 1976–1980. *Eur J Oncol* 1999; 4:645–651.

17. Wolff MS, Schecter A. Accidental exposure of children to polychlorinated biphenyls. *Arch Environ Contam Toxicol* 1991; 20:449–453.

18. Brock JW, Melnyk LJ, Caudill SP, Needham LL, Bond AR. Serum levels of several organochlorine pesticides in farmers correspond with dietary exposure and local use history. *Toxicol Ind Health* 1998; 14:275–289.

19. Beck H, Eckart K, Mathar W, Witt-Kowski L. Levels of PCDD's and PCDF's in adipose tissue of occupationally exposed workers. *Chemosphere* 1989; 18:507–516.

20. Schecter A, Papke O, Ball M, Lis A, Brandt-Rauf P. Dioxin concentrations in the blood of workers at municipal waste incinerators. *Occup Environ Med* 1995; 52:385–387.

21. Kang HK, Watanabe KK, Breen JJ, Remmers J, Conomos MG, Stanley J, et al. Dioxins and dibenzofurans in adipose tissue of US Vietnam veterans and controls. *Am J Public Health* 1991; 81:344–349.

22. Dewailly E, Ayotte P, Bruneau S, Laliberte C, Muir DCG, Norstrom RJ. Inuit exposure to organochlorines through the food chain in Arctic Quebec. *Environ Health Perspect* 1993; 101:618–620.

23. Mes J, Marchand L, Davies DJ. Organochlorine residues in adipose tissue of Canadians. *Bull Environ Contamin Toxicol* 1990; 45:681–688.

24. Ryan JJ, Schecter A, Masuda Y, Kikuchi M. Comparison of PCDDs and PCDFs in the tissue of Yusho patients with those from the general population in Japan and China. *Chemosphere* 1987; 16:2017–2025.

25. Murphy R, Harvey C. Residues and metabolites of selected persistent halogenated hydrocarbons in blood specimens from a general population survey. *Environ Health Perspect* 1985; 60:115–120.

26. Stehr-Green PA. Demographic and seasonal influences on human serum pesticide residue levels. *J Toxicol Environ Health* 1989; 27:405–421.

27. Savage E, Keefe TJ, Tessari JD, Wheeler HW, Applehans FM, Goes EA, et al. National study of chlorinated hydrocarbon insecticide residues in human milk, USA. I. Geographic distribution of dieldrin, heptachlor, heptachlor epoxide, chlordane, oxychlordane, and mirex. *Am J Epidemiol* 1981; 113:413–422.

28. Orban JE, Stanley JS, Schwemberger JG, Remmers JC. Dioxins and dibenzofurans in adipose tissue of the general US population and selected subpopulations. *Am J Public Health* 1994; 84(3):439–445.

29. Office of Prevention, Pesticides and Toxic Substances. *NHAT comparability study*. 1992 Jun. EPA #700-R–12–010.

30. Stanley JS, Ayling RE, Cramer PH, Thornburg KR, Remmers JC, Breen JJ, et al. Polychlorinated dibenzo-p-dioxin and dibenzofuran concentration levels in human adipose tissue samples from the continental United States collected from 1971 through 1987. *Chemosphere* 1990; 20:895–901.

31. Wallace LA, Pellizzari ED, Hartwell TD, Whitmore R, Zelon H, Perritt R, et al. The California TEAM study: Breath concentrations and personal exposures to 26 volatile compounds in air and drinking water of 188 residents of Los Angeles, Antioch, and Pittsburg, CA. *Atmos Environ* 1988; 22:2141–2163.

32. Pellizzari ED, Hartwell TD, Harris BS, Waddell RD, Whitaker DA, Erickson MD. Purgeable organic compounds in mother's milk. *Bull Environ Contam Toxicol* 1982; 28:322–328.

33. Centers for Disease Control and Prevention. *National report on human exposure to environmental chemicals*. Available: http://www.cdc.gov/nceh/dls/report

34. Hill RH, Head S, Baker S, Gregg M, Shealy DB, Bailey SL, et al. Pesticide residues in urine of adults living in the United States: Reference range concentrations. *Environ Res* 1995; 71:99–108.

35. Pirkle JL, Needham LL, Sexton K. Improving exposure assessment by monitoring human tissues for toxic chemicals. *J Exp Anal Environ Epidemiol* 1995; 5:405–424.

36. Paschal DC, Ting BG, Morrow JC, Pirkle JL, Jackson RJ, Sampson EJ, et al. Trace metals in urine of United States residents: Reference range concentrations. *Environ Res* 1998; 76:53–59.

37. Needham LL, Patterson DG Jr, Burse VW, Paschal DC, Turner WE. Reference range data for assessing exposure to selected environmental toxicants. *Toxicol Ind Health* 1996; 12:507–513.

38. Needham LL, Hill RH, Ashley DL, Pirkle JJ, Sampson EJ. The priority toxicant reference range study: Interim report. *Environ Health Perspect* 1995; 103:89–94.

39. Leder A, Linak E, Holmes R, Sasano T. *CEH marketing research report: Chlorine/sodium hydroxide*. Palo Alto, CA: SRI International, 1994.

40. United Nations Environment Programme. *Decisions adopted by the Governing Council at its nineteenth session: 13C. International action to protect human health and the environment through measures which will reduce and/or eliminate emissions and discharges of persistent organic pollutants, including the development of an international legally binding instrument*. Geneva: United Nations, 1997.

41. Great Lakes Water Quality Board. *1987 report on the Great Lakes water quality*. Windsor, ON: International Joint Commission, 1987.

42. American Public Health Association. Resolution 9304: Recognizing and addressing the environmental and occupational health problems posed by chlorinated organic chemicals. *Am J Public Health* 1994 84:514–515.

43. Dewailly E, Mulvad G, Pedersen HS, Ayotte P, Demers A, Weber J-P, et al. Concentration of organochlorines in human brain, liver, and adipose tissue autopsy samples from Greenland. *Environ Health Perspect* 1999; 107:823–828.

44. MacKay D. Is chlorine really the evil element? *Environ Sci Eng* 1992 Nov:49–56.

45. Webster T, Commoner B. Overview: The dioxin debate. In: Schecter A, editor. *Dioxins and health*. New York: Plenum Press, 1994.

46. Colborn T, vom Saal FS, Soto AM. Developmental effects of endocrine-disrupting chemicals in wildlife and humans. *Environ Health Perspect* 1993; 101:378–384.

47. Gray LE, Ostby JS, Kelce WR. A dose-response analysis of the reproductive effects of a single gestational dose of 2,3,7,8-tetrachlorodibenzo-p-dioxin in male Long Evans hooded rat offspring. *Toxicol Appl Pharmacol* 1997; 146:11–20.

48. Guillette LJ, Crain DA. *Endocrine disruption: An evolutionary perspective*. London: Taylor and Francis, 2000.

49. US Environmental Protection Agency. *Estimating exposures to dioxin-like compounds*. Vol. 1–3, review draft. Washington, DC: US Environmental Protection Agency, Office of Research and Development, 1994. EPA/600/6–88–005.

50. Backer LC, Ashley DL, Bonin MA, Cardinali FL, Kieszak SM, Wooten JV. Household exposures to drinking water disinfection by-products: Whole blood trihalomethane levels. *J Expo Anal Environ Epidemiol*. 2000; 10:321–326.

51. Porter WP, Jaeger JW, Carlson IH. Endocrine, immune, and behavioral effects of aldicarb (carbamate), atrazine (triazine), and nitrate (fertilizer) mixtures at groundwater concentrations. *Toxicol Ind Health* 1999 15:133–150.

52. Wu C, Schaum J. Exposure assessment of trichloroethylene. *Environ Health Perspect* 2000; 108 Suppl 2:359–363.

53. Hill RH Jr, Ashley DL, Head SL, Needham LL, Pirkle JL. p-Dichlorobenzene exposure among 1,000 adults in the United States. *Arch Environ Health* 1995; 50:277–280.

54. Tichenor BA, Sparks LE, Jackson MD, Guo Z, Mason M, Plyunket CM, et al. Emissions of perchloroethylene from dry cleaned fabrics. *Atmos Environ* 1990; 24A:1219–1229.

55. Zartarian VG, Ozkaynak H, Burke JM, Zufall MJ, Rigas ML, Furtaw EJ Jr. A modeling framework for estimating children's residential exposure and dose to chlorpyrifos via dermal residue contact and nondietary ingestion. *Environ Health Perspect* 2000; 108:505–514.

56. Minh TB, Watanabe M, Tanabe S, Yamada T, Hata J, Watanabe S. Occurrence of tris(4-chloropheyl)methane, tris(4-chlorophenyl)methanol, and some other persistent organochlorines in Japanese human adipose tissues. *Environ Health Perspect* 2000; 108:599–603.

57. Lippman M, editor. *Environmental toxicants: Human exposures and their health effects*. 2nd ed. New York: Wiley Interscience, 2000.

58. Ashley DL, Bonin MA, Cardinali FL, McGraw JM, Wooten JV. Measurement of volatile organic compounds in human blood. *Environ Health Perspect* 1996; 104:871–877.

59. Kutz FW, Wood PH, Bottimore DP. Organochlorine pesticides and polychlorinated biphenyls in human adipose tissue. *Rev Environ Contam Toxicol* 1991; 120:1–82.

60. Blount BC, Silva MJ, Caudill SP, Needham LL, Pirkle JL, Sampson EJ, et al. Levels of seven urinary phthalate metabolites in a human reference population. *Environ Health Perspect* 2000; 108:972–982.

61. Ligon WV, Dorn SB, May RJ. Chlorodibenzofuran and chlorodibenzo-p-dioxin levels in Chilean mummies dated to about 2800 years before the present. *Environ Sci Technol* 1989; 23:1286–1290.

62. Schecter A, Dekin A, Weerasinghe NCA, Arghestani S, Gross ML. Sources of dioxins in the environment: A study of PCDDs and PCDFs in ancient frozen Eskimo tissue. *Chemosphere* 1988; 17:627–631.

63. Tong HY, Gross MLO, Schecter A, Monson SJ, Dekin A. Sources of dioxin in the environment: Second stage study of PCDD/Fs in ancient human tissue and environmental samples. *Chemosphere* 1990; 20:987–992.

64. Schecter A. Exposure assessment: Measurement of dioxins and related chemicals in human tissues. In: Schecter A, editor. *Dioxins and health*. New York: Plenum Press, 1994.

65. Ryan JJ, Schecter A, Lizotte R, Sun W-F, Miller L. Tissue distribution of dioxins and furans in humans from the general population. *Chemosphere* 1985; 14:929–932.

66. Andrews JS Jr, Garrett WA Jr, Patterson DG Jr, Needham LL, Roberts DW, Bagby JR, et al. 2,3,7,8-tetrachlorodibenzo-p-dioxin levels in adipose tissue of persons with no known exposure and in exposed persons. *Chemosphere* 1989; 18:499–506.

67. US Environmental Protection Agency, National Center for Environmental Assessment. *Exposure and human health reassessment of 2,3,7,8-tetrachlorodibenzo-p-dioxin (TCDD) and related compounds, Part III* (SAB review draft). Washington, DC: US Environmental Protection Agency, 2000.

68. Pirkle JL, Wolff WH, Patterson DG, Needham LL, Michalek JE, Miner JC, et al. Estimates of the half-life of 2,3,7,8-tetrachlorodibenzo-p-dioxin in Vietnam veterans of Operation Ranch Hand. *J Toxicol Environ Health* 1989; 27:165–171.

69. LaKinjd JS, Berlin CM, Naiman DQ. Infant exposure to chemicals in breast milk in the United States: What we need to learn from a breast milk monitoring program. *Environ Health Perspect* 2001; 109:75–88.

70. Tatsukawa R, Tanabe S. Fate and bioaccumulation of persistent organochlorine compounds in the marine environment. In: Baumgartner DJ, Dudall IM, editors. *Oceanic processes in marine pollution*. Vol. 6. Malabar, FL: RE Krieger, 1990:39–55.

71. Arctic Monitoring and Assessment Programme. *Arctic pollution issues: A state of the Arctic environment report*. Oslo: AMAP Directorate, 1997.

72. Gilman A, Dewailly E, Feeley M, Jerome V, Kuhnlein H, Kwavnick B, et al. Human health. In: Jensen J, Adare K, Shearer R, editors. *Canadian Arctic contaminants assessment report*. Ottawa: Indian and Northern Affairs Canada, 1997.

73. Phillips LJ. A comparison of human toxics exposure and environmental contamination by census division. *Arch Environ Contam. Toxicol* 1992; 22:1–5.

74. Phillips LJ, Birchard G. Regional variations in human toxic exposure in the USA: An analysis based on the National Adipose Tissue Survey. *Arch Environ Contam Toxicol* 1991; 21:159–168.

75. Patterson DG Jr, Holtman RE, Needham LL, Roberts DW, Bagby JR, Pirkle JL, et al. 2,3,7,8-tetrachlorodibenzo-p-dioxin levels in adipose tissue of exposed and control persons in Missouri. *JAMA* 1986; 256:2683–2686.

76. US Centers for Disease Control. *Health consultation: Exposure investigation report, Calcasieu Estuary, Lake Charles, Calcasieu Parish, Louisiana*. Washington, DC: Department of Health and Human Services, 1999.

77. Cocco P, Kazerouni N, Hoar Zahm S. Cancer mortality and environmental exposure to DDE in the United States. *Environ Health Perspect* 2000; 108:1–4.

78. Kutz FW, Yobs AR, Strassman SC. Racial stratification of organochlorine insecticide residues in human adipose tissue. *J Occup Med* 1977; 19:619–622.

79. Agency for Toxic Substances and Disease Registry. *The nature and extent of lead poisoning in children in the United States: A report to Congress.* Washington, DC: US Department of Health and Human Services, 1988.

80. Mahaffey KR. National estimates of blood lead levels: United States, 1976–1980: Association with selected demographic and socioeconomic factors. *New Engl J Med* 1982; 307:573–579.

81. Brown P. Race, class, and environmental health. *Environ Res* 1995; 69:15–20.

82. Mossing ML, Redetzke KA, Applegate HG. Organochlorine pesticides in blood of persons from El Paso, Texas. *J Environ Health* 1985; 47:312–313.

83. Mes J. Temporal changes in some chlorinated hydrocarbon residue levels of Canadian breast milk and infant exposure. *Environ Pollution* 1994; 84:261–268.

84. Alcock RE, Jones KC. Chemical contaminants in human milk. *Environ Sci Technol* 1996; 30:3133–3143.

85. Schettler T, Solomon G, Valenti M, Huddle A. *Generations at risk: Reproductive health and the environment.* Cambridge, MA: MIT Press, 1999.

86. Jensen AA. Chemical contaminants in human milk. *Residue Rev* 1983: 89:1–128.

87. Koppe JG, Pluim HJ, Olie K, van Wijnen J. Breast milk, dioxins, and the possible effects on the health of newborn infants. *Sci Total Environ* 1991; 106:33–41.

88. Anderson HA, Wolff MS. Environmental contaminants in human milk. *J Exp Anal Environ Epidemiol* 2000; 10:1–6.

89. Bern HA. The fragile fetus. In: Colborn T, Clement C, editors. *Chemically-induced alterations in sexual and functional development: The wildlife/human connection.* Princeton, NJ: Princeton Scientific Publishing, 1992. pp. 1–8.

90. Schecter A, Papke O, Ball M. Evidence for transplacental transfer of dioxins from mother to fetus: Chlorinated dioxin and dibenzofuran levels in the livers of stillborn infants. *Chemosphere* 1990; 21:1017–1022.

91. Schecter A, Ryand JJ, Papke O. Decrease in levels and body burden of dioxins, dibenzofurans, PCBs, DDE, and HCB in blood and milk in a mother nursing twins over a thirty-eight month period. *Chemosphere* 1998; 37:1807–1816.

92. Jacobson JL, Jacobson SW. Intellectual impairment in children exposed to polychlorinated biphenyls in utero. *N Engl J Med* 1996; 335:783–789.

93. Foster W, Chan S, Platt L, Hughes C. Detection of endocrine disrupting chemicals in samples of second trimester human amniotic fluid. *J Clin Endocrinol Metab* 2000; 85:2954–2957.

94. Schecter A, Gasiewicz T. Health hazard assessment of chlorinated dioxins and dibenzofurans contained in human milk. *Chemosphere* 1987; 16:2147–2154.

95. Furst P. PCDD/PCDFs in human milk—still a matter of concern? *Organohalogen Compounds* 2000; 48:111–114.

96. Berkowitz T. The rationale for a national prospective cohort study of environmental exposure and childhood development. *Environ Res* (Section A) 2001; 85:59–68.

97. Anderson LM, Beebe LE, Fox SD, Issaq HJ, Kovatch RM. Promotion of mouse lung tumors by bioaccumulated polychlorinated aromatic hydrocarbons. *Exp Lung Res* 1991; 17:455–471.

98. Juberg DR. Traces of environmental chemicals in the human body: Are they a risk to health? *Am Counc Sci Health* 1999 Apr; 1–15.

99. Bailar JC. How dangerous is dioxin? *N Engl J Med* 1991; 324:260–262.

100. Rose G. Environmental factors and disease: The man-made environment. *Br Med J* 1987; 294:963–965.

101. Tryphonas H. Immunotoxicity of PCBs (Aroclors) in relation to Great Lakes. *Environ Health Perspect* 1995; 103 Suppl 9:35–46.

102. McKinney JD, Pedersen LG. Do residue levels of polychlorinated biphenyls (PCBs) in human blood produce mild hypothyroidism? *J Theor Biol* 1987; 129:231–241.

103. Ulrich EM, Caperell-Grant A, Jung S, Hites RA, Bigsby RM. Environmentally relevant xenoestrogen tissue concentrations correlated to biological responses in mice. *Environ Health Perspect* 2000; 108:973–977.

104. Longnecker MP, Rogan WJ, Lucier G. The human health effects of DDT (dichlorodiphenyl-trichloroethane) and PCBs (polychlorinated biphenyls) and an overview of organochlorines in public health. *Annu Rev Public Health* 1997; 18:211–244.

105. Goldman LR, Harnly M, Flattery J, Patterson DG, Jr. Needham LL. Serum polychlorinated dibenzo-p-dioxins and polychlorinated dibenzofurans among people eating contaminated home-produced eggs and beef. *Environ Health Perspect* 2000; 108:13–19.

106. Jacobson JL, Jacobson SW, Humphrey HEB. Effects of in utero exosure to polychlorinated biphenyls and related contaminants on cognitive functioning in young children. *J Pediatr* 1990; 116:38–45.

107. Jacobson JL, Jacobson SW. A 4-year followup study of children born to consumers of Lake Michigan fish. *J Great Lakes Res* 1993, 19:776–783.

108. Gladen BC, Rogan WJ, Hardy P, Thullen J, Tingelstad J, Tully M. Development after exposure to polychlorinated biphenyls and dichlorodiphenyl dichloroethene transplacentally and through human milk. *J Pediatr* 1988; 113:991–995.

109. Gladen BC, Rogan WJ. Effects of perinatal polychlorinated biphenyls and dichlorodiphenyl dichloroethene on later development. *J Pediatr* 1991; 119:58–63.

110. Rogan WJ, Gladen BC, McKinney JD, Carreras N, Hardy P, Thullen J, et al. Neonatal effects of transplacental exposure to PCBs and DDE. *J Pediatr* 1986; 109:335–341.

111. Rogan WJ, Gladen BC. PCBs, DDE, and child development at 18 and 24 months. *Ann Epidemiol* 1991; 1:407–413.

112. Patandin S, Dagnele PC, Mulder PGH, Op de Coul E, van der Veen JE, Weisglas-Kuperus N, et al. Dietary exposure to polychlorinated biphenyls and dioxins from infancy until adulthood: A comparison between breast-feeding, toddler, and long-term exposure. *Environ Health Perspect* 1999; 107:45–51.

113. Huisman M, Koopman-Esseboom C, Fidler V, Hadders-Algra M, Van der Paauw CG, Tuinstra LG, et al. Perinatal exposure to polychlorinated biphenyls and dioxins and its effect on neonatal neurological development. *Early Hum Dev* 1995; 41:111–127.

114. Huisman M, Koopman-Esseboom C, Lanting CI, Van der Paauw CG, Tuinstra LG, Fidler V, et al. Neurological condition in 18-month-old children perinatally exposed to polychlorinated biphenyls and dioxins. *Early Hum Dev* 1995; 43:165–176.

115. Koopman-Esseboom C, Weisglas-Kuperus N, de Ridder MA, Van der Paauw CG, Tuinstra LG, Sauer PJ. Effects of polychlorinated biphenyl/dioxin exposure and feeding type on infants' mental and psychomotor development. *Pediatrics.* 1996; 97:700–706.

116. Patandin S, Koopman-Esseboom C, de Ridder MA, Weisglas-Kuperus N, Sauer PJ. Effects of environmental exposure to polychlorinated biphenyls and dioxins on birth size and growth in Dutch children. *Pediatr Res* 1998; 44:538–545.

117. Pluim HJ, de Vijlder JJM, Olie K, Kok JH, Vulsma T, van Tijn DA, et al. Effects of pre- and postnatal exposure to chlorinated dioxins and furans on human neonatal thyroid hormone concentrations. *Environ Health Perspect* 1993; 101:504–508.

118. Rylander L, Stromberg U, Dyremark E, Ostman C, Nilsson-Ehle P, Hagmar L. Polychlorinated biphenyls in blood plasma among Swedish fish consumers in relation to low birth weight. *Am J Epidemiol* 1998; 147:493–502.

119. Rylander L, Stromberg U, Hagmar L. Decreased birthweight in infants born to women with a high dietary intake of fish contaminated with persistent organochlorine compounds. *Scand J Work Environ Health* 1995; 21:368–375.

120. Rylander L, Stromberg U, Hagmar L. Dietary intake of fish contaminated with persistent organochlorine compounds in relation to low birthweight. *Scand J Work Environ Health* 1996; 22:260–266.

121. Koppe JG, Pluim E, Olie K. Breastmilk, PCBs, dioxins and vitamin K: A discussion paper. *J R Soc Med* 1989; 82:416–419.

122. Koppe JG, Pluim HJ, Olie K, van Wijnen J. Breast milk, dioxins and the possible effects on the health of newborn infants. *Sci Total Environ* 1991; 106:33–41.

123. Dewailly E, Dodin S, Verreault R, Ayotte P, Sauve L, Morin J. High organochlorine body burden in women with estrogen-receptor positive breast cancer. *J Natl Cancer Inst* 1994; 86:232–234.

124. Djordevic MV, Hoffman D, Fan J, Prokopczyk B, Citron ML, Stellman SD. Assessment of chlorinated pesticides and polychlorinated biphenyls in adipose breast tissue using a supercritical fluid extraction method. *Carcinogenesis* 1994; 15:2581–2585.

125. Falck F, Ricci A, Wolff MS, Godbold J, Deckers P. Pesticides and polychlorinated biphenyl residues in human breast lipids and their relation to breast cancer. *Arch Environ Health* 1992; 47:143–146.

126. Hoyer AP, Grandjean P, Jorgensen T, Brock JW, Hartvig HB. Organochlorine exposure and risk of breast cancer. *Lancet* 1998; 352:1816–1829.

127. Moysich KB, Ambrosone CB, Vena JE, Shields PG, Mendola P, Kostnyiak P, et al. Environmental organochlorine exposure and postmenopausal breast cancer risk. *Cancer Epidemiol Biomarkers Prev* 1998; 7:181–188.

128. Mussalo-Rahmaa H, Hasanen E, Pyysalo H, Kauppila AR, Pantzar P. Occurrence of beta-hexachlorocyclohexane in breast cancer patients. *Cancer* 1990; 66:2124–2128.

129. Rylander L, Hamgar L. Mortality and cancer incidence among women with a high consumption of fatty fish contaminated with persistent organochlorine compounds. *Scand J Work Environ Health* 1995; 21:419–426.

130. Wolff MS, Toniolo PG, Lee EW, Rivera M, Dubin N. Blood levels of organochlorine residues and risk of breast cancer. *J Natl Cancer Inst* 1993; 85:648–652.

131. Hardell L, Lindstrom G, van Bavel B, Fredrikson M, Liljegren G. Some aspects of the etiology of non-Hodgkin's lymphoma. *Environ Health Perspect* 1998; 106 (supp 2):679–681.

132. Hardell L, van Bavel B, Lindstrom G, Fredrikson M, Hagberg H, Liljegren G, et al. Higher concentrations of specific polychlorinated biphenyl congeners in adipose tissue from non-Hodgkin's lymphoma patients compared with controls without a malignant disease. *Int J Oncol* 1996; 9:603–608.

133. Rothmann N, Cantor KP, Blair A, Bush D, Brock JW, Helzlouer K, et al. A nested case-control study of non-Hodgkin's lymphoma and serum organochlorine residues. *Lancet* 1997; 350:240–244.

134. Svensson BG, Mikoczy Z, Stromberg U, Hagmar L. Mortality and cancer incidence among Swedish fishermen with a high dietary intake of persistent organochlorine compounds. *Scand J Work Environ Health* 1995; 21:106–115.

135. Hunter DJ, Hankinson SE, Laden F, Colditz GA, Manson JE, Willett WC, Speizer FE, Wolff MS. Plasma organochlorine levels and the risk of breast cancer. *N Engl J Med* 1997; 337:1253–1258.

136. Krieger N, Wolff MS, Hiatt RA, Rivera M, Vogelman J, Orentreich N. Breast cancer and serum organochlorines: a Prospective study among white, black, and Asian women. *J Natl Cancer Inst* 1994; 86:589–599.

137. van't Veer P, Lobbezo IE, Martin-Moreno JM, Guallar E, Gomez-Aracena J, Kardinal AFM, et al. DDT (dicophane) and postmenopausal breast cancer in Europe: Case-control study. *Br Med J* 1997; 315:81–85.

138. Osius N, Karmaus W, Kruse H, Witten J. Exposure to polychlorinated biphenyls and levels of thyroid hormones in children. *Environ Health Perspect* 1999; 107:843–849.

139. Koopman-Esseboom C, Morse D, Weisglas-Kuperus, N, Lutkeschipholt IJ, Van der Paauw CG, Tuinstra LG, et al. Effects of dioxins and polychlorinated biphenyls on thyroid hormone status of pregnant women and their infants. *Pediatr Res* 1994; 36:468–473.

140. Pluim HJ, Koppek G, Olie I. Effects of dioxins and furans on thyroid hormone regulation in the human newborn. *Chemosphere* 1993; 27:391–394.

Cancer and the Environment

Richard W. Clapp

11

Workplace and community exposures to toxic materials and how they influence the risk of cancer have been among the urgent environmental health issues of the past quarter century. In the United States, much of the concern has focused on chemicals and radiation, both ionizing and nonionizing, and their relationship to clusters of cancer in communities, factories, and sometimes schools. Citizens' and workers' organizations have focused on some dramatic examples, such as Times Beach, Missouri,[1] and Love Canal, New York,[2] which were both evacuated because of community exposure to carcinogenic chemicals, such as dioxin, and the Oak Ridge National Laboratory in Tennessee,[3] where excess radiation has increased the risk of leukemia deaths in the workforce. In Europe, widespread concern followed the chemical plant explosion in Seveso, Italy,[4] and the Chernobyl, Ukraine, nuclear-plant disaster.[5] But numerous less publicized examples have occurred in communities throughout North America, and health care providers have been asked to evaluate the causes and implications of many such local concerns. In this chapter, three aspects of the problem are addressed: (1) the background of cancer incidence and mortality against which local clusters or excess cases are assessed, (2) the types of exposures that are known or suspected causes of such clusters, and (3) the implications for health care providers who wish to provide guidance for worried patients or communities.

Understanding Cancer Trends

In the past quarter century, the incidence of cancer in the United States has risen steadily, from an age-adjusted rate for all sites combined of 320.0 per 100,000 in 1973 to 395.0 per 100,000 in 1998.[6] In Canada, the age-adjusted incidence rate (using the Canadian population distribution as the standard) rose from 280.2 and 338.3 per 100,000 in 1972 for females and males, respectively, to an estimated 343.9 and 444.5 per 100,000 in 2001.[7] The age-adjusted cancer mortality rates in the two countries have risen and recently fallen during this time period. Because of improved treatments for some cancers and declining lung cancer mortality in males, the all-sites cancer mortality rate in 1975 was quite similar to the mortality rate in 1997 in both countries. But within the overall rates, some important trends in specific types of cancer are revealing.

Table 11.1 lists the major types of cancer for which there are significant trends in incidence, mortality, or both. Some of these trends, such as the decreasing incidence and mortality from stomach cancer, are not new and have been associated with improved food quality and safety over the past century. Likewise, the decreasing incidence and mortality due to cancer of the cervix is most likely the result of intensive screening efforts and therapeutic efficacy. The most worrisome trends, however, are the cancer

Table 11.1 Trends in U.S. cancer incidence and mortality for selected sites, 1973–1998

Sites with decreasing incidence and mortality	Sites with increasing incidence and mortality
Oral cavity and pharynx	Esophagus
Stomach	Liver and intrahepatic bile ducts
Colon and rectum	Lung and bronchus
Pancreas	Melanomas of skin
Larynx	Prostate
Cervix uteri	Kidney and renal pelvis
Corpus and uterus, not otherwise specified	Brain and other nervous system
Hodgkin's disease	Non-Hodgkin's lymphoma
Leukemias	Multiple myeloma

Source: Adapted from reference 6.

types in which both incidence and mortality have been increasing for the past two decades or more. These include lung cancer, especially in females, non-Hodgkin's lymphoma, melanoma of the skin, multiple myeloma, and several other less common malignancies. Some of these—for example, non-Hodgkin's lymphoma—have been associated with environmental and occupational exposures to carcinogens.

Recently, attention in the United States has focused on declining mortality rates for some of the more common cancer types, including breast and gynecological cancer in females, prostate and lung cancer in males, and colorectal cancer in both sexes. This trend was particularly evident during 1991–1995, when there was a 2.6 percent decline in the age-adjusted cancer death rate. Some observers attributed this to reduced smoking in males and better treatment of the more common cancers other than lung cancer.[8] While any reversal in the long-term upward trend in cancer mortality rates is welcome, the recent decline is really quite modest. Considering the decrease in mortality due to heart disease and stroke over the past 25 years and the enormous resources devoted toward reducing mortality from cancer during the same time period, it is surprising that more progress has not been made.

Table 11.2 shows the different trends in cancer mortality in those under age 65 and those aged 65 and older in the United States.[6] The recent

Table 11.2 49-Year trends in U.S. cancer mortality rates

Age group	1950	1975	1998	% change 1950–1998
0–4	11.1	5.2	2.3	−77.5
5–14	6.6	4.7	2.6	−59.6
15–24	8.5	6.6	4.5	−48.2
25–34	19.8	14.6	10.9	−43.6
35–44	64.2	53.9	38.9	−38.4
45–54	175.2	179.2	134.2	−22.6
55–64	394.0	423.3	385.9	−0.7
65–74	700.0	769.8	830.2	19.4
75–84	1160.9	1156.0	1320.3	14.7
85+	1450.7	1437.9	1751.4	21.5

Source: Adapted from reference 6.

Note: All races, males and females; All primary cancer sites combined.

improvements in cancer mortality are clearly in those younger than 65, where less than half the cancer deaths occur. The overall age-adjusted rates, which have been standardized to the 1970 U.S. population structure, tend to "overweight" these deaths in younger individuals. As the convention shifts to calculating rates adjusted to the 2000 U.S. population, these trends will change somewhat, and the declines will appear less pronounced. This means that the ranking of cancer deaths relative to heart disease may eventually reverse, and cancer may become the leading cause of death in the country.

Another measure of the national burden of cancer is the estimated lifetime risk of being diagnosed with an invasive malignancy. The most recent published data from the National Cancer Institute SEER calculate this risk as 43 percent for males and 38 percent for females.[6] In Canada, the corresponding lifetime risks are 40 percent for males and 36 percent for females.[7] These are higher than the often-quoted "one in three" and partly reflect the overall aging of the North American population. However, these lifetime risks carry with them the prospect of much suffering and distress for cancer patients and their families—an incentive to look for avoidable causes of cancer, along with the ongoing effort to improve treatment.

Occupational and Environmental Causes of Cancer

Several authors have attempted to quantify the avoidable causes of cancer. In 1981, a widely cited report by Doll and Peto identified tobacco products and diet as the largest contributors to cancer mortality in U.S. whites under age 65, with a relatively small contribution by occupation and industrial pollution.[9] More recently, a group of authors at the Harvard Center for Cancer Prevention revised these estimates slightly and suggested that 30 percent of total cancer deaths were due to tobacco and another 30 percent were due to adult diet/obesity.[10] These authors estimate that 5 percent of cancer deaths are due to occupational factors, 2 percent to environmental pollution, and the remainder to a variety of other factors. The accuracy of these estimates is open to question, but it is not productive to play one cancer cause off against another. Just because an avoidable cause of cancer does not have the large impact that tobacco has on the overall cancer burden, it should not be neglected. Clearly, some factors interact and magnify the effect of each acting separately. Two examples of such synergistic exposures are asbestos and tobacco smoke in workplaces and tobacco smoke and radon in homes. The point is to take action wherever possible to reduce cancer risk.

Two cancer types that have been increasing rapidly in recent years are melanoma of the skin and non-Hodgkin's lymphoma. The causes of these two cancers include environmental and occupational exposures to ultraviolet light,[11] as well as genetic factors and, in the case of non-Hodgkin's lymphoma, phenoxyacetic acid herbicides, solvents, and viral exposures.[12] Table 11.3 lists other occupational and environmental exposures that the

Table 11.3 Some established occupational and environmental carcinogens

Exposure	Target organ in humans
Aflatoxins	Liver
4-aminobiphenyl	Bladder
Arsenic	Lung, skin
Asbestos	Lung, pleura, peritoneum
Benzene	Hematopoietic system
Benzidine	Bladder
Beryllium	Lung
Bis(chloromethyl)ether	Lung
Cadmium	Lung
Coal-tar pitches	Skin, lung, bladder
Erionite	Pleura
Mineral oils	Skin
Mustard gas	Pharynx, lung
2-naphthylamine	Bladder
Nickel compounds	Nasal cavity, lung
Radon	Lung
Shale oils	Skin
Silica	Lung
Solar radiation	Skin
Soots	Skin, lung
Talc containing asbestiform fibers	Lung
2,3,7,8-tetrachloro-dibenzo-p-dioxin	Lung, soft-tissue sarcoma
Tobacco smoke	Lung, bladder, oral cavity, pharynx, larynx, esophagus
Vinyl chloride	Liver, lung, blood vessels

Source: References 13–15.

International Agency for Research on Cancer (IARC) considers as causing cancer in humans (Group 1 in the IARC classification system).

Over the past 25 years, the list of established human carcinogens has grown substantially. This is due, in part, to the accumulation of knowledge about the human health effects of various chemicals, drugs, and other exposures during this period. Other exposures currently being evaluated include electric and magnetic fields and environmental estrogens. The latter two are of particular concern because of the size of the population that may be exposed in industrialized countries.

Implications for Cancer Prevention

Most concerns about environmental causes of cancer have arisen because of widely publicized events, such as toxic pollution of communities around industrial sites or nuclear facilities. Many of these events are localized, and the exposed population may be limited to the workers within the industries or the communities immediately around the facilities. Typically, the exposures to workers are greater than to residents of surrounding communities, but there are unusual events in which community exposures have been substantial. Community exposures are also of concern because, in addition to working adults, residents include infants, children, and the elderly. Exposure of newborns to radioactive iodine, for example, may increase the subsequent risk of thyroid cancer more than equivalent exposure to adults.[13] This is the reason that environmental regulations are more stringent than workplace regulations for the same substance.

While recent downward trends in cancer mortality are welcome, there are ongoing environmental and occupational exposures in many communities and workplaces that could be avoided. Indeed, despite the phasing out of stratospheric ozone-depleting chemicals over the past decade, the amount of ultraviolet radiation reaching the earth's surface will continue to increase into the next century. (See chapter 8.) Dioxin (2,3,7,8-tetrachlorodibenzo-p-dioxin) in the workplace environment has decreased as a result of the banning of herbicides like 2,4,5-T in North America. But dioxin is still produced during incineration of solid waste that contains PVC plastic. The U.S. Environmental Protection Agency has identified medical-waste incineration as the second-leading source of emissions of dioxin and dioxin-like compounds in the United States.[14] This is emblematic of an environmental problem that has direct implications for human cancer and that should be a concern for all health care providers. Rather than attempting to diminish these problems, or rank them below other causes of cancer, it is in everyone's interest to take them seriously and seek opportunities to prevent further exposure.

The usual recommendations for cancer prevention, such as those offered by the American Cancer Society,[15] include smoking cessation, eating a diet high in plant foods and low in meat and dairy products, increasing physical activity, and getting periodic checkups and screening tests. While these are rational recommendations, they are necessarily general and not focused on particular circumstances about which exposed workers or community residents may be concerned.

Health care providers who wish to provide additional advice and help for such patients or communities need to inquire about what the exposures have been. Public databases such as the U.S. EPA Toxics Release Inventory,[16] the Canadian National Pollutant Release Inventory,[17] or local agency records can provide some insight into preventable exposures. A careful exposure history[18] or an understanding of toxic releases from industrial facilities can provide the basis for targeted cancer prevention that will be helpful to a broader population.

Environmental Exposure Assessment

Persons may be exposed to hazardous materials in the home, workplace, school, or other (e.g., travel or recreational) settings. Assessment should consider all potential, not just occupational, exposures. In many cases, persons may have significant exposures without symptoms or recognition of hazard. An exposure history has several components, including a description of the patient's past and present jobs, a listing of materials used in the home, yard, garden, garage or shop, and any exposure at school, travel, or play. Does the patient live with a person exposed to chemicals or radiation in the workplace? Has the patient ever lived in proximity to industrial plants or used contaminated drinking water? If an exposure is identified, then information about duration of exposure and the nature and name of the materials is appropriate, including the timing of any symptoms in relation to exposure. Dietary history may suggest exposure, as in sports or subsistence fish eaters exposed to mercury and PCBs. Smoking, alcohol, and use of recreational and medical drugs are also important in this assessment. As in most clinical matters, the more a physician knows about the life experience of the "whole" patient, the more likely that physician is to identify relevant hazardous exposures.[18]

Conclusion

Cancer rates are a measure of the burden of this disease in populations. They represent the aggregate of many devastating problems for individual patients and their families. The causes of cancer are still under investigation, but we know enough about many causes to advocate for prevention of exposures wherever possible. Advocates for worker health and safety and grassroots environmental organizations have taken the lead role in the past two or three decades. Health care providers can support this ongoing effort by working with others to prevent exposure to carcinogenic chemicals, in addition to advising patients about a healthy lifestyle.

References

1. Gerrard MB. *Whose backyard, whose risk?* Cambridge, MA: MIT Press, 1994.

2. Levine AG. *Love Canal: Science, politics, and people.* Lexington, MA: Lexington Books, 1982.

3. Wing S, Shy CM, Wood JL, Cragle DL, Frome EL. Mortality among workers at Oak Ridge National Laboratory: Evidence of radiation effects in follow-up through 1984. *JAMA* 1991; 265:1397–1402.

4. Bertazzi PA, Zocchetti C, Guercilena S, Consonni D, Tironi A, Landi MT, et al. Dioxin exposure and cancer risk: A 15-year mortality study after the "Seveso Accident." *Epidemiol* 1997; 8:646–652.

5. Bard D, Verger P, Hubert P. Chernobyl, 10 years after: Health consequences. *Epidemiol Rev* 1997; 19(2):187–204.

6. Ries LAG, Kosary CL, Hankey BF, Harras A, Miller BA, Edwards BK, editors. *SEER cancer statistics review, 1973–1998.* Bethesda, MD: National Cancer Institute, 2001.

7. *Canadian cancer statistics, 1970–2001.* Ottawa: Health Canada, 2001. Available: http://www.cancer.ca/stats/trendse.htm [Graphs of trends for specific cancer types for males and females available at http://www.cancer.ca/stats/fig2ae.htm]

8. Cole P, Radu B. Declining cancer mortality in the United States. *Cancer* 1996; 78:2045–2048.

9. Doll R, Peto R. The causes of cancer: Quantitative estimates of avoidable risks of cancer in the United States. *J Natl Cancer Inst* 1981; 66:1191–1308.

10. Harvard Center for Cancer Prevention. Harvard Report on Cancer Prevention. *Cancer Causes Control* 1996; 7:S1–S59.

11. Gilchrest BA, Eller MS, Geller AC, Yaar M. The pathogenesis of melanoma induced by ultraviolet light. *New Engl J Med* 1999; 340:1341–1348.

12. Scherr PA, Mueller NE. Non-Hodgkin's lymphomas. In: Schottenfeld D, Fraumeni JF, editors. *Cancer epidemiology and prevention*. 2nd ed. New York: Oxford University Press, 1996.

13. Gilbert ES, Tarone R, Bouville A, Rom E. Thyroid cancer rates and I-131 doses from Nevada atmospheric nuclear bomb tests. *J Natl Cancer Inst* 1998; 90:1654–1660.

14. Environmental Protection Agency. *Exposure and human health reassessment of 2,3,7,8-tetrachlorodibenzo-p-dioxin (TCDD). Part III: Integrated summary and risk characterization for 2,3,7,8-tetrachlorodibenzo-p-dioxin (TCDD) and related compounds.* 2000 Sept. EPA/600/P–00/001Bg. Available: http://www.epa.gov/ncea

15. American Cancer Society. Atlanta: American Cancer Society, 2000. Available: http://www3.cancer.org/cancerinfo/

16. Available: http://www.rtk.net

17. Available: http://www.ec.gc.ca/pdb/npri/

18. ATSDR case studies in environmental medicine no. 26. Taking an exposure history. In: Pope AM, Rall DP, editors. *Environmental medicine*. Washington, DC: National Academy Press, 1995.

Radiation and Health

Maureen Hatch and Michael McCally

12

The public has long been concerned about the danger to health from exposure to ionizing radiation. The concern has been particularly keen when exposure is generated by nuclear-power-plant accidents or by contamination from nuclear facilities during normal operations. At least three major nuclear-reactor accidents have occurred since the advent of nuclear power: one at Windscale (now Sellafield), England, in 1957; one at the Three Mile Island plant in Pennsylvania in 1979; and one at Chernobyl in the former Soviet Union in 1986. Other "significant" radiation accidents have occurred at military, research, and commercial facilities.[1] It is less well appreciated that radon gas emanating from underlying geological formations or building materials in homes is a far more common and significant source of public exposure to ionizing radiation than emissions from nuclear plants (figure 12.1).

The biological science of how radiation affects the body is well understood, but the epidemiologic science that tries to relate very low radiation exposures to subsequent cancer and other illness is based on still unproven—some say unprovable—hypotheses. This lack of scientific certainty fuels the public controversy over what level of radiation exposure is acceptable.

In recent years, nonionizing radiation has captured public attention. Suspicion has fallen chiefly on the extremely low frequency fields emitted by electric power lines and household appliances, such as electric blankets. Cellular telephones, which fall in the higher-frequency, microwave portion of the electromagnetic spectrum, have also been a target of concern (figure 12.2).

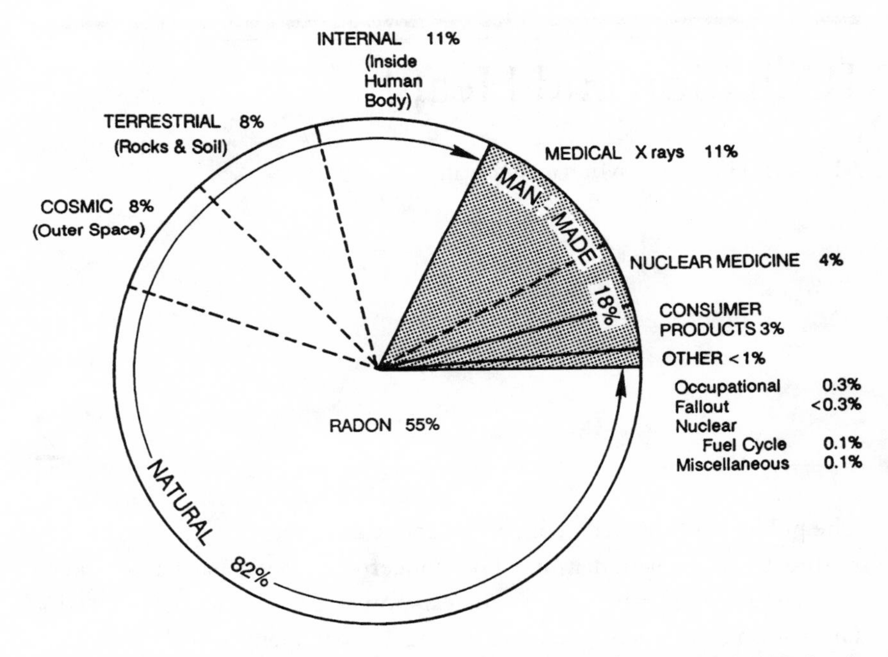

Figure 12.1 Sources of human exposure to ionizing radiation. *Source:* National Academy of Science. *Health Effects of Exposure to Low Levels of Ionizing Radiation Beir V*. Washington, DC: National Academy Press; 1990

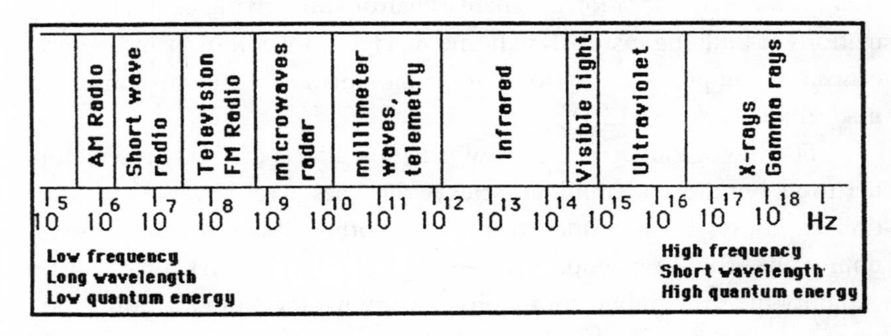

Figure 12.2 The electromagnetic spectrum. *Source:* http://hyperphysics.phy-astr.gsu.edu/hbase/ems1.html

But public perceptions of the risk of nuclear technology depend only in part on scientific evidence. Kai Erikson's interviews with nearby residents after the accident at Three Mile Island document the special dread that radiation inspires.[2] Many radiation scientists complain that people do not understand that smoking and driving are much more dangerous than exposure to low levels of radiation. And some blame the media for misrepresenting the Three Mile Island accident, in which the officially reported release of radioactive iodine (I-131) was quite small (13–20 curies, versus 50 million curies at Chernobyl).[3] Erikson points out that, for the public, radiation's invisibility and its lingering effects make it unlike other dangers.[2] Public policy must take this special fear into account.

Types of Radiation

The electromagnetic spectrum is divided into three segments with visible light in the middle. The other two segments are referred to as *ionizing* and *nonionizing,* based on their biological activity, and span a wide range of wavelengths. The ionizing portion of the spectrum includes gamma rays and ultraviolet radiation. Ionizing radiation is given off by decaying radioisotopes, or radionuclides, typically as beta particles or gamma rays. (While alpha and beta particles and cosmic gamma rays are ionizing, they are not electromagnetic and hence are not shown in figure 12.2.) The distinguishing feature of ionizing radiation is that it has a photon or particle sufficiently energetic to break chemical bonds, liberating electrons and damaging the DNA of an atom or molecule in tissue either directly or indirectly through the formation of free radicals. Thus, ionizing radiation is genotoxic and can act as a carcinogen.

Nonionizing radiation is not energetic enough to disrupt electrons and hence is not thought to be genotoxic. The radio-frequency range of the nonionizing-radiation spectrum can heat tissue. Below that, in the range of low and extremely low frequency (ELF) fields with their tremendously long wavelengths, very little biological activity has been noted. It is these fields that are associated with power (50–60 Hertz (Hz)) lines, as well as with household appliances.

Although electromagnetic fields (EMFs) were initially believed to be biologically inactive, some mechanisms have been proposed to link EMFs to adverse health effects.[4] For example, the so-called melatonin hypothesis postulates a reduction in the pineal gland's nocturnal production of melatonin, which in turn could increase levels of estrogen and prolactin, reduce melatonin's inhibitory effect on cell proliferation, and increase susceptibility to DNA damage through a reduction in melatonin's antioxidant action. Evidence for the melatonin effect of EMFs in both animal and human studies has been mixed. Effects on cell signaling and on the immune system have also been proposed. Mechanistic investigations are important to this area of research since biological plausibility is all but essential if associations between an exposure and outcome are to carry any weight.

Health Effects of Ionizing Radiation

What we know about radiation health effects comes principally from three sources: (1) long-term follow-up of the survivors of the atomic bombings in Hiroshima and Nagasaki, which involved an acute exposure to high radiation doses at high dose rates; (2) studies of patients receiving diagnostic and therapeutic radiation, also high dose but typically intermittent; and (3) studies of occupational cohorts, typically with protracted exposures. Of these three, Atomic Bomb Survivor data have served as the basis for extrapolating downward to exposure at low doses and low dose rates, and for setting radiation-protection standards (see table 12.1).

Ionizing radiation produces two kinds of cell injury, one immediate and the other delayed. High-level doses—a term loosely designating exposures over 100 rem—inevitably produce the kind of immediate, direct

Table 12.1 Nomenclatures and units* for radiation exposure and dose

Quantity	SI unit	Conventional unit (symbol and value in SI units)
Exposure	No special name	1 roentgen [R] = 2.58×10^{-4} C kg^{-1}
Absorbed dose	1 gray [Gy] = 1 J kg^{-1}	1 rad [rad] = 0.01 Gy
Dose equivalent	1 sievert [Sv]† = 1 J kg^{-1}	1 rem [rem]† = 0.01 Sv

Source: National Council on Radiation Protection, 1989.
*Where the abbreviations *C*, *kg* and *J* in the table stand for coulomb, kilogram, and joule, respectively.
†The magnitude of the dose equivalent in sievert or rem depends on the LET of the radiation.

effects seen in the Chernobyl firefighters: skin burns, hair loss, bone-marrow destruction, and damage to the intestinal lining. The severity of the Chernobyl workers' injuries, some of which were fatal, was directly related to the amount of radiation received. High radiation doses are also carcinogenic. The effects of low doses—a loose term for exposures under 10 rem—are not immediately visible and are far less predictable because they involve the cancerous transformation of cells. The complex process that produces cancerous tumors has many intermediate steps, and the changes may happen in some persons and not in others. It may take years for epidemiologists, using statistical methods, to establish an increased incidence of cancer in the population of exposed persons (for example, uranium miners who received low doses of radiation by inhaling radon gas and other radioisotopes).[5,6]

Most, but not all, scientists believe that such stochastic or probabilistic effects of radiation are also directly related to the radiation dose and that they can occur at any dose, no matter how small. This is called the *linear, no-threshold hypothesis;* it means that all exposure to radiation presents some risk to human health. This hypothesis is still controversial. Until recently, most radiation scientists felt that, because cells repaired themselves, no association with cancer would be found at radiation doses below a certain threshold. The National Academy of Sciences report on the Biological Effects of Ionizing Radiation (BEIR-V), published in 1990, accepts the linear, no-threshold hypothesis, a significant change from earlier scientific opinion.[7]

Although the relation between very low doses of radiation and cancer is still unresolved, there is little argument about the effects of high and moderately high doses of radiation.[1,8] The direct, immediate effects of radiation are seen in all cases, from the highest doses down to about 100 rem. Scientists also agree that persons exposed to levels of 10–50 rem have an increased likelihood of developing cancer. These conclusions are based on experiences with x-rays, as well as on observations of bomb and accident victims. Early radiologists, unaware of potential harm from the new Roentgen rays (x-rays), suffered both direct effects (skin burns and hair loss) and indirect effects (higher rates of leukemia some years later). In Great Britain and the United States in the 1930s, several thousand children were treated for scalp ringworm with high doses (200–400 rem) of x-rays. Other patients received local irradiation for spinal disease and benign gynecological disorders, including breast inflammation. All these groups later were shown to have increased rates of leukemia and certain types of cancer.[9]

Exposure Standards

BEIR-V raised the estimate of cancer risk associated with low-level radiation exposure from estimates in earlier BEIR reports. In the 1980s, scientists observed more cancer deaths among the bomb victims than had been predicted. At the same time, improved calculations lowered the estimates of the gamma radiation released by the bombs. In other words, less radiation had produced more cancer. And, as mentioned above, investigators have abandoned the notion of a threshold below which no cancer effect would be observed.

In response to these findings, the International Commission on Radiological Protection has set the exposure allowed for nuclear workers, medical personnel, and others in contact with radiation at 5 rem per year. The standards for occupational and public radiation exposure have become progressively more stringent over the last 60 years.[9] In 1987, new information on radon gas in households led the National Council on Radiation Protection and Measurement to more than double its estimate of the average American's background radiation exposure, from 170 millirem per year to 360 millirem. That is equivalent to about 18 chest x-ray film exposures, and it is still small relative to the 5-rem annual exposure that the federal government allows in the workplace.[9]

Atomic Bomb Survivors

Because cancer is common, one must study very large numbers of persons to explore specific causes—otherwise the incidence of cancer in an irradiated group would not be measurably different from the natural incidence. There are very few large populations that have been exposed to known levels of radiation. The 106,000 surviving victims of Hiroshima and Nagasaki (Atomic Bomb Survivors or ABS), are the best known and studied. The doses they received have been painstakingly reconstructed as part of the work of a large, ongoing research program set up in 1946—the

Atomic Bomb Casualty Commission, now known as the Radiation Effects Research Foundation (RERF), based in Hiroshima.[10,11]

Close follow-up of the survivors of the 1945 atomic bombings in Hiroshima and Nagasaki has demonstrated an excess of most cancers, with leukemia (except for chronic lymphatic leukemia), thyroid, and breast appearing to be the most radiosensitive sites.[12] A striking finding (replicated in certain of the non-Hiroshima/Nagasaki medical cohorts) is that the highest radiation-related increase in breast cancer is in women less than age 10 at the time of the bombing.[13] Higher risks have also been seen for leukemia and thyroid cancer following radiation exposures in childhood compared to exposures at later ages.

For many years, the ABS cohort has served as the gold standard for all radiation health studies, in spite of the fact that the exposure was to a single high dose.[14] Findings from populations exposed to low doses for protracted periods of time have been challenged if they differed from the radiobiological predictions based on the ABS data. It should be noted, however, that despite the differences in exposure, the ABS cohort has recently provided evidence of risk for doses as low as 0.1 Sv (10 rem), although this is still below the range of interest for the general population.[15]

Workers exposed to radiation have also been studied. Occupational cohorts can provide reasonably accurate individual doses because of their use of personal dosimetry. They have the added benefit of allowing direct investigation of low-dose effects. However, sample size has been a serious limitation of some worker-cohort studies, with results often conflicting or indeterminate.[10,11] A recent approach involves examining the data from nationwide registries to increase the number of study subjects.[16]

Epidemiologists' Problem with Cancer Studies

Epidemiologists investigate the causes of disease in groups rather than in individual patients. Proving causation is the most difficult part of their task, and it relies on judgment. Low-level radiation is associated with increased cancer incidence, but association is not causation. Scientists test associations for strength, consistency, specificity, and plausibility before making judgments about causation. For example, lung cancer is more common in smokers than in nonsmokers, and the consistent judgment of most scientists, after the usual tests, is that smoking "causes" cancer.

Fallout from Nuclear Testing

Effects of radioactive fallout in southwestern Utah from the Nevada Test Site detonations in the 1950s were studied by Stevens et al., using a sophisticated model for dose estimation that predicted the pattern of radioactive deposition and took account of each individual's intake of cow's milk and locally grown foods.[17] The study reported a striking sevenfold increase in deaths from leukemia in those 19 years of age. A later report[18] suggested a link between thyroid neoplasm's and exposure to I-131. No controversy surrounds the investigations conducted on the Marshall Islanders who were exposed to both I-131 and gamma radiation during the BRAVO test in 1954. Exposure led to increases in thyroid cancer and benign thyroid nodules. Disturbances in growth were also noted among exposed children.[19]

Accidents at Nuclear Plants

The accident at the Three Mile Island (TMI) nuclear plant in Pennsylvania in March 1979 was unprecedented in the United States and roused international attention and great anxiety in the surrounding area. In 1986, one of us (MH) and colleagues undertook a study evaluating cancer among residents of a 10-mile radius with respect to releases of radiation during the accident. To estimate exposure, we used a mathematical dispersion model, which accounted for modifying factors such as weather, wind, and terrain and which was validated against readings from off-site dosimeters. Although we found raised risks of some cancers, the pattern of results was not convincing that emissions from the plant had influenced cancer risk during the limited (seven-year) follow-up period.[20] A study published subsequently based on longer follow-up also found no substantial evidence of adverse effects.[21] A trend in breast cancer with gamma-radiation exposure subsequently disappeared after longer follow-up. (See Wing et al. for a critique of these studies and their interpretation.)[22,23]

Exposure from the accident at TMI was low—estimated at less than 0.5 mSv (50 mrem) for those within five miles of the plant, or less than the annual dose from background radiation. However, the 1986 accident at Chernobyl in the former Soviet Union was another matter. Radioactive contamination is now widespread in Belarus, Russia, and the Ukraine, and Chernobyl fallout has affected much of Northern Europe. The general population of the contaminated areas in the former Soviet Union is estimated to have received a dose of 5–10 mSv. Those who have continued living there have an estimated lifetime dose of 5.0 mSv. The accident recovery workers received very high doses, on the order of 100 mSv (10 rem).[23]

Beginning in 1992, a relatively few years after the Chernobyl accident, there have been reports of an extremely large increase in thyroid cancer in children, particularly in those under age four at the time of the accident. The current yield of cases is far greater than what was predicted by models based on other populations exposed to radioactivity in childhood, particularly for I-131. Improvements in the number of cases detected by screening have been suggested as a possible explanation but seems unlikely to account for the entire increase. Several studies now in progress should help to clarify the picture, but there is no discounting the large number of cases or their early onset following exposure. While formal studies of the accident recovery workers at Chernobyl are still in progress, increases in leukemia in this population are already being reported.[23]

Routine Nuclear-Power-Plant Operations

There have been several reports of increases in leukemia among children living near certain nuclear facilities in Great Britain, although radiobiological models would not predict any detectable increase in cancer risk from the low levels of radioactivity emitted by nuclear installations.[24] A follow-up study related the excess of childhood cancer near the Sellafield nuclear facility, which has had some previous accidents, to preconception exposure of fathers working in the plant.[25] However, this novel hypothesis involving paternal radiation exposure prior to conception has not been confirmed in later studies.[26]

Environmental Contamination from Nuclear Plants

Only in 1986 did the U.S. Department of Energy reveal that the Hanford facility in Richland, Washington, had been releasing radioactive material, primarily I-131, into the environment over a period of years. A study was launched to examine exposures to the thyroid and rates of thyroid disease.[27] Exposure estimation was derived from the model used in the Hanford Environmental Dose Reconstruction Project. It accounted for a great many factors: first, factors that would determine how much I-131 would stay on vegetation, and then factors affecting contamination of milk and, ultimately, consumption, and deposition in the thyroid gland. Residents were screened using ultrasound to detect subtle thyroid abnormalities, in addition to thyroid nodules and thyroid cancer. Results from this study are expected to be published shortly. An initial report suggests no radiation effect.[27]

Nuclear-Waste Containment

A special health hazard associated with nuclear power is the production of vast inventories of radioactive wastes. These wastes consist of uranium mill tailings, transuranic wastes (defined as materials containing more than 100 nanocuries of transuranic material per gram of waste), "high-level wastes" (which include spent fuel rodes and the sludges left behind when spent fuel rods are reprocessed to extract plutonium), and "low-level wastes" (which include wastes not included in the other categories). "Low-level wastes" may include a wide range of materials, many of which are intensely radioactive. A typical 55-gallon drum of medical "low-level" waste will contain on the order of 10 millicuries. A typical drum of wastewater from a nuclear reactor may contain 10 curies, and a drum of irradiated reactor components may contain 200–1,000 curies.

Commerical power reactors produce the vast majority of all "low-level" radioactive wastes, whether measured by volume or by total curies. In the area of "high-level wastes," the commercial nuclear-power industry in the United States alone will have produced 41,000 metric tons by the year 2000. There is no established safe technology for dealing with these wastes.

Low-level wastes have been stored at a number of waste dumps. The sites that have been studied most extensively are those associated with the nuclear-weapons program. All of these have experienced major breaches of containment, with significant (in some cases, massive) contamination of surrounding areas (e.g., Hanford). Current federal law requires states, acting individually or in cooperation with other states, to build a new generation of "low-level" waste-site dumps. In the United States, spent fuel rods are currently stored in cooling pools at reactor sites, pending creation of an acceptable repository for them. No reported leaks from these storage pools have been reported, but there have been massive leaks from storage facilities for high-level waste from the nuclear-weapons program.[28]

One unique aspect of the danger associated with radioactive wastes derives not only from the extreme toxicity of these materials, but also from the great longevity of many of their isotopes. Cesium–137 and strontium–90 have half-lives of 30 and 28 years, respectively, and must be isolated for 300–600 years. Plutonium–239 has a half-life of 24,400 years and must be isolated for 240,000–480,000 years. The United Nations has been in existence only since 1945, and it has failed to prevent widespread civil and international wars even during this brief period. Violent social and political upheaval is common. (All of recorded human history spans only 6,000 years.) We continue to produce huge quantities of highly toxic radioactive materials, knowing that to safeguard these materials would require an intact and functioning social order over thousands of years.

Radon

As figure 12.1 shows, radon represents over half of all the ionizing radiation exposures of the general population. Radon is a gas and can accumulate to high levels in enclosed spaces, such as underground mines and the basements of residences. The decay products of radon, called *radon progeny,* attach to dust particles. Once inhaled, further decay occurs and can expose the lining of the lung to alpha radiation, potentially initiating a carcinogenic process.[29] Although inhalation of radon gas from stone and soil is the main source of exposure, off-gassing of radon from water can also add to cumulative radon exposures.

Initially, the estimates of risk of radon-related lung cancer were derived by extrapolating downward from studies of underground miners.[30] More recent studies have been addressing population levels of residential radon directly.[31] Both types of study have found evidence of excess risk at the higher residential-exposure levels. Whether the risk posed by radon varies with smoking status is an open question. Earlier studies of miners suggested that cigarette smoking increased lung cancer risk in persons exposed to radon,[32] but a recent study in Sweden of residential radon did not find appreciably different relative risks in smokers and never-smokers.[33] However, they did report that environmental tobacco smoke may increase radon-related lung cancer risk among never-smokers.

Health Effects of Nonionizing Radiation

Electromagnetic radiation does not produce ionization products, and energy transfer to human tissue during exposure is low. As a result, health effects are subtle and difficult to measure.

Power Lines

The initial suggestion of an adverse health effect from the 50–60 Hz alternating fields associated with household electricity came in studies linking an increase in childhood cancer with residential proximity to power lines.[34] Because of their ubiquity, electric and magnetic fields in the power frequency range (60 Hz) of commercial electrical power have attracted the attention of researchers from many countries.[35] The National Cancer Institute has carried out a nationwide study of power lines and childhood leukemia, specifically acute lymphocytic leukemia (ALL), the most common type in children.[36] The study staff measured magnetic fields in the child's bedroom and other rooms of the house. The investigators found no association of cancer rates with either the direct measurements of magnetic fields or with surrogate measures. Because of the conflicting results from many studies, a symposium was convened in 1998 by the National Institute of Environmental Health Sciences to provide a forum for review, discussion, and consensus statements about whether causal relations were likely. The experts in attendance agreed that a modest elevation in risk of childhood leukemia was present for higher exposure categories, but did not feel that a causal relationship had been demonstrated.[37] Serious limitations of the current studies include the absence of an agreed-on exposure metric and the lack of a biological basis for the hypothesis that EMFs either cause or promote cancer in children. It should be noted, however, that in spite of initial skepticism by biophysicists that nonionizing fields are too weak to cause tissue heating or other biological effects, the existence of "nonthermal" effects such as interference with cell signaling and calcium homeostasis are now fairly well accepted. A developing viewpoint holds that if EMFs are carcinogenic, it is as a promoter, not an initiator.

Household Appliances

Radiation exposure from the use of electric blankets—particularly during pregnancy—has been assessed in relation to a number of adverse reproductive endpoints, as well as childhood cancer. There is at least one report linking maternal use of electric blankets during pregnancy with birth weight and fetal development, but the number of exposed cases was too small to produce a stable result.[38] Findings with respect to other reproductive endpoints are mixed.[39,40]

Occupational groups exposed to ELF, such as electrical workers, have also been studied for a number of different cancers, principally brain cancer and leukemia. Two meta-analyses combining the data from these studies found no overall evidence that exposure to magnetic fields was causally linked to risk of leukemia or brain cancer.[41,42] Recently, an experienced researcher in the field has suggested that these cohorts are simply too small to provide unequivocal results for such rare cancers.[43]

Cellular (Mobile) Telephones

The rapid increase in use of cellular telephones has, as with any ubiquitous exposure, begun to attract the attention of researchers. Public concern arose following a well-publicized report on brain cancer in a frequent user, on the same side of his head on which he used the phone.[44] The signals these phones transmit and receive fall in the microwave part of the electromagnetic spectrum, so again, there is no obvious mechanism to link them to brain cancer. Only a few studies have so far been conducted. A recent nationwide cancer-incidence study in Denmark did not support the hypothesis of an association between use of mobile phones and tumors of the brain, salivary gland, leukemia, or other cancers.[45] Other recent studies are consistent with these results.[46,47]

Prescriptions

The most frequent human radiation exposure is to radon. Fortunately, it can be easily measured and easily mitigated—essentially by increasing ventilation in enclosed spaces where radon accumulates. Avoiding building materials that generate radon gas or housing sites on geological formations associated with radon would obviate the problem almost entirely.

Exposure to nonionizing radiation is also common: power lines, mobile phones, and, to a lesser extent, electric blankets are ubiquitous. While there is not yet a complete consensus in the scientific community, large or even moderate health effects from these exposures seem unlikely at this point. However, EMF exposure measurement continues to be a serious barrier to producing unequivocal results. Until further information develops, the best course is prudent avoidance.

In the case of cellular phones, simply use them less frequently or with a headset. In terms of ionizing radiation, exposure of the general population to I-131 is particularly significant because of the pasture-grass-cow-milk-thyroid pathway, which can result in substantial doses to the thyroid gland. Sophisticated methods have been used to estimate individual exposure through this pathway (described above). Young children are in general more sensitive to radiation-induced thyroid disease than older children or adults. The instances of I-131 exposure that have received attention have been atypical: radioactive fallout from aboveground testing, releases of material from the Hanford facility, and the accident at Chernobyl. It is not easy to propose policy strategies to guard against a repeat of such occurrences. Sweden and Germany have chosen to completely phase out their nuclear-power industries. While some perhaps may support this strong "antinuclear" position, such a policy is unlikely in the United States at present. However, the recommendation has been made in some localities to stockpile potassium iodine pills that would be administered to suppress I-131 uptake by the thyroid gland following any future reactor accident.

In the event I-131 releases do occur, there is limited comfort in the fact that survival from thyroid cancer is quite good and that removal of the thyroid can be managed with subsequent thyroid replacement therapy. Careful attention to the topics selected for future research is clearly required. A number of important studies are now underway. These include the Chernobyl investigations of workers and the general population. This new generation of studies will test the adequacy of radiation-protection guidelines for workers, the general public, and sensitive subgroups. The high incidence and early onset of thyroid cancer in young children in the Chernobyl area points to a higher sensitivity than was evident from the ABS and medical cohorts.[48] Well-informed studies with good underlying mechanistic hypotheses will provide the best basis for health-protective radiation-exposure policy.

Editor's Note

As significant as the above-mentioned threats are to public health, they pale in comparison with the greatest danger posed by nuclear power: its central role in the proliferation—actual and potential—of nuclear weapons. India and Pakistan now have nuclear weapons. Fears about Iraq's and North Korea's attempts to develop nuclear weapons have further underlined the dangers of nuclear proliferation.

The critical role of nuclear power in weapons proliferation has not received adequate attention, even though this role has been obvious from the outset of the nuclear era. In 1946, the Acheson-Lilienthal report warned that "the development of atomic energy for peaceful purposes and the development of atomic energy for bombs are in much of their course interchangeable and interdependent."[49]

Highly enriched uranium, one of the two fissionable materials that can be used to make nuclear weapons, is not used in commercial reactors but is used in "peaceful" research reactors. Iraq's nuclear threat derives from its possession of 100 pounds of highly enriched uranium—enough to make two bombs equal in yield to the bomb that destroyed Nagasaki.[50] Plutonium, the other fuel for nuclear weapons, is used directly in some commercial power plants. The growth of the plutonium-reprocessing industry will lead to the production of vast quantities of this substance. Over the next 30 years, Japan alone could produce for its nuclear-power industry 400 metric tons of plutonium—twice the amount currently in the combined arsenals of all the nuclear-weapons powers.[49] Such inventories of plutonium will create extraordinary opportunities for diversion to non-nuclear states intent on weapons of development and to terrorist and criminal organizations.

As the Acheson-Lilienthal report predicted in 1946, the "peaceful" nuclear-power industry has provided the opportunity for would-be nuclear powers to acquire both the weapons-grade fuel and the cadre of skilled scientific and technical personnel needed to make atomic bombs.[49] It is difficult to imagine how we can hope to prevent the runaway proliferation of nuclear weapons as long as the nuclear-power industry continues to produce raw materials usable for weapons.

It is proposed that fuel rods from commercial reactors ultimately be stored in deep geological repositories, where, it is hoped, the waste will be safely isolated from the environment. However, no functioning repositories now exist. The proposed repository sites at Yucca Mountain in Nevada have been criticized for failing to meet the criteria for geological stability. The projected opening date of this facility keeps receding as new problems and political oppositions are encountered.

References

1. Mettler F Jr, Ricks RC. Historical aspects of radiation accidents. In: Mettler FA, Ricks CA, Ricks RC, editors. *Medical management of radiation accidents.* CRC Press, 1990.

2. Erikson K. Toxic reckoning: Business faces a new kind of fear. *Harvard Business Review* 1990 Jan-Feb. pp. 118–126.

3. U.S. Department of Energy. *Health and environmental consequences of accidents.* Washington, DC: US Department of Energy, 1987. DOE Report DOE/ER–0332.

4. EMF Rapid (Electric and Magnetic Fields Research and Public Information Dissemination Program). *EMF Science Review Symposium Breakout Group Reports, 1998: Mechanisms.* Research Triangle Park, NC: National Institute of Environmental Health Science, 1998. Available: www.niehs.nih.gov/emfrapid/home.htm

5. Wagoner JK, Archer VE, Carroll BE. Cancer mortality patterns among United States uranium miners and millers, 1950–1962. *J Cancer Inst* 1964; 32:787–801.

6. Kahn PA. Grisly archive of key cancer data. *Science* 1993; 259:448–451.

7. National Academy of Sciences. *BEIR-V: Health effects of exposure to low levels of ionizing radiation.* Washington, DC: National Academy Press, 1990.

8. Biological effects of low-dose radiation: A workshop. *Health Physics* 1990; 59(1): 1–102.

9. Caufield C. *Mutiple exposure: Chronicles of the radiation age.* New York: Harper & Row, 1989.

10. Mancuso TF, Stewart AM, Kneale G. Radiation exposures of Hanford workers dying from cancer and other causes. *Health Phys* 1997; 33:369–385.

11. Stewart AM, Kneale GW. An overview of the Hanford controversy. *Occup Med* 1991; 6: 641–663. Review.

12. Schull WJ. Radioepidemiology of the A-bomb survivors. *Health Phys* 1996; 70:798–803.

13. Land CE, Tokunaga M, Nakamura N. Early-onset breast cancer among Japanese A-bomb survivors: Evidence of a radiation susceptible subgroup? *Lancet* 1993; 342:237.

14. Gilbert ES. Invited commentary: Studies of workers exposed to low doses of radiation. *Am J Epidemiol* 2001; 153:319–322.

15. Pierce DA, Preston DL. Radiation-related cancer risks at low doses among atomic bomb survivors. *Radiat Res* 2000; 154:178–186.

16. Sont WN, Zielinski JM, Ashmore JP, Jiang H, Krewski D, Fair ME, et al. First analysis of cancer incidence and occupational radiation exposure based on the National Dose Registry of Canada. *Amer J Epidemiol* 2001; 153:309–318.

17. Stevens W, Thomas DC, Lyon JL, Till JE, Kerber RA, Simon SL, et al. Leukemia in Utah and radioactive fallout from the Nevada Test Site. *JAMA* 1990; 264:585–591.

18. Kerber RA, Till JE, Simon SL, Lyon JL, Thomas DC, Preston-Martin S, et al. A cohort study of thyroid disease in relation to fallout from nuclear weapons testing. *JAMA* 1993; 270:2076–2082.

19. Takahashi T, Trott KR, Fujimori K, Simon SL, Ohtomo H, Nakashima N, et al. An investigation into the prevalence of thyroid disease on Kwajalein Atoll, Marshall Islands. *Health Phys* 1997; 73:199–213.

20. Hatch MC, Beyea J, Nieves JW, Susser M. Cancer near the Three Mile Island nuclear plant: Radiation emissions. *Am J Epidemiol* 1990; 132:397–417.

21. Talbott EO, Youk AO, McHugh KP, Shire JD, Zhang A, Murphy BP, et al. Mortality among the residents of the Three Mile Island accident area: 1979–1992. *Environ Health Persp* 2000; 108:545–552.

22. Wing S, Richardson D. Collision of evidence and assumptions: TMI deja view. *Environ Health Persp* 2000; 108:A546–A547.

23. Cardis E, Richardson D, Kesminiene A. Radiation risk estimates in the beginning of the 21st century. *Health Physics* 2001; 80:349–361.

24. Hatch M. Childhood leukemia around nuclear facilities. *Sci Total Environ* 1992; 127:37–42.

25. Gardner MW, Snee MP, Hall AJ. Results of case-control study of leukemia and lymphoma among young people near Sellafield nuclear plant. *Br Med J* 1990; 300:423–429.

26. Parker L, Craft AW, Smith J, Dickinson H, Wakerford R, Binks K, et al. Geographical distribution of preconceptional radiation doses to fathers employed at the Sellafield nuclear installation, West Cumbria. *Br Med J* 1993; 307:966–971.

27. National Research Council. *Review of the Hanford Thyroid Disease Study: Draft, final report.* Washington, DC: National Academy Press, 2000.

28. Department of Energy, Oak Ridge National Laboratory. *Spent fuel and radioactive waste inventories: Projections and characteristics.* Oak Ridge, TN: Department of Energy, Oak Ridge National Laboratory, 1986. Report DOW/RW–0006.

29. National Research Council. Committee on Biological Effects of Ionizing Radiation (BEIR-V). *Health effects of exposure to low levels of ionizing radiation.* Washington, DC: National Academy Press, 1990.

30. Howe GR, Nair RC, Newcombe HB, Miller AB, Abbatt JD. Lung cancer mortality (1950–1980) in relation to radon daughter exposure in a cohort of workers at the Eldorado Beaverlodge uranium mine. *J Natl Cancer Inst* 1986; 77:357–362.

31. Schoenberg JB, Klotz JB, Wilcox HB, Nicholls GP, Gil-del-Real MT, Stemhagen A, et al. Case-control study of residential radon and lung cancer among New Jersey women. *Cancer Res* 1990; 50:6520–6524.

32. National Research Council, Committee on Biological Effects of Ionizing Radiation (BEIR-IV). *Health effects of exposure to radon.* Washington, DC: National Academy Press, 1988.

33. Lagarde F, Axelsson G, Damber L, Mellander H, Nyberg F, Pershagen G. Residential radon and lung cancer among never-smokers in Sweden. *Epidemiology* 2001; 12:396–404.

34. Wertheimer N, Leeper E. Electrical wiring configurations and childhood cancer. *Am J Epidemiol* 1979; 109:273–284.

35. Feychting M, Ahlbom A. Magnetic fields and cancer in children residing near Swedish high-voltage power lines. *Am J Epidemiol* 1993; 138:467–481.

36. Linet MS, Hatch EE, Kleinerman RA, Robison LL, Kaune WT, Friedman DR, et al. Residential exposure to magnetic fields and acute lymphoblastic leukemia in children. *N Engl J Med* 1997; 337:1–7.

37. EMF Rapid (Electric and Magnetic Fields Research and Public Information Dissemination Program). *EMF Science Review Symposium Breakout Group Reports, 1998: Childhood cancer.* Research Triangle Park, NC: National Institute of Environmental Health Sciences, 1998. Available: www.niehs.nih.gov/emfrapid/home.htm.

38. Bracken MB, Belanger K, Hellenbrand K, Dlugosz L, Holford TR, McSharry JE, et al. Exposure to electromagnetic fields during pregnancy with emphasis on electrically heated beds: Association with birthweight and intrauterine growth retardation. *Epidemiology* 1995; 6:263–270.

39. Hatch M. The Epidemiology of electric and magnetic field exposures in the power frequency range and reproductive outcomes. *Ped and Perin Epidmiol* 1992; 6:198–214.

40. EMF Rapid (Electric and Magnetic Fields Research and Public Information Dissemination Program). *EMF Science Review Symposium Breakout Group Reports, 1998: Reproductive outcomes.* Research Triangle Park, NC: National Institute of Environmental Health Sciences, 1998. Available: www.niehs.nih.gov/emfrapid/home.htm

41. Kheifets LI, Afifi AA, Buffler PA, Zhang ZW. Occupational electric and magnetic field exposure and brain cancer: A meta-analysis. *J Occup Environ Med* 1995; 37:1327–1341.

42. National Institute of Environmental Health Sciences. *NIEHS report on health effects from exposure to power line frequency: Electric and Magnetic fields.* Research Triangle Park, NC: National Institute of Environmental Health Sciences, 1999. NIH Publication No. 99–4493. Available: http://www.niehs.nih.gov/emfrapid/home.htm.

43. Savitz DA. Invited commentary: Electromagnetic fields and cancer in railway workers. *Am J Epidemiol* 2001; 153:836–838.

44. Park RL. Cellular telephones and cancer: How should science respond? *J Natl Cancer Inst* 2001; 93:166–167.

45. Johansen C, Boice JD, McLaughlin JK, Olsen JH. Cellular telephones and cancer—A nationwide cohort study in Denmark. *J Natl Cancer Inst* 2001; 93:203–207.

46. Muscat JE, Malkin MG, Thompson S, Shore RE, Stellman SD, McRee D, et al. Handheld cellular telephone use and risk of brain cancer. *JAMA* 2000; 284:3001–3007.

47. Inskip PD, Tarone RE, Hatch EE, Wilcosky TC, Shapiro WR, Selker RG, et al. Cellular telephone use and brain tumors. *N Engl J Med* 2001; 344:79–86.

48. Ron E, Lubin JH, Shore RE, Mabuchi K, Modan B, Pottern LM, et al. Thyroid cancer after exposure to external radiation: A pooled analysis of seven studies. *Radiat Res* 1995; 141:259–277.

49. Department of State. *Committee of Atomic Energy report on the international control of atomic energy (the "Acheson-Lillenthal" report).* Washington, DC: Department of State, 1946. p. 4.

50. Leventhal PL. *Latent and blatant proliferation: Does the NPT work against either?* Washington, DC: Nuclear Control Institute, 1990.

The Science of Risk Assessment

John C. Bailar III and A. John Bailer

13

Potential hazards surround us at home, in the workplace, in our cars, and even in health care facilities. Given exposure to these hazards, we may want to evaluate the risk that an adverse event will occur or that it will occur at some level of severity. In this chapter, we introduce some concepts about risk and how it can be assessed, comment on the nature of hazards and the uncertainties inherent in the risk-assessment process, and show how risk assessment affects the management of these hazards. Most of our discussion is in terms of chemical carcinogenesis, but the principles apply to the full range of threats to human health and survival.

Scope of Risk Assessment

In 1983, a committee of the U.S. National Academy of Sciences proposed a model that is now commonly used to discuss the assessment of occupational and environmental hazards.[1] Other individuals and agencies, such as the National Research Council,[2] have elaborated and refined this model. The basic model, which we use here, partitions risk assessment into four steps: hazard identification, dose-response modeling, exposure assessment, and risk characterization. Integration of a risk assessment with a cost analysis and other matters to develop strategies for risk regulation and control is often called *risk management*.

Numerous scientific and technical disciplines are involved throughout a risk assessment. Hazard identification uses the input of biologists, chemists, and others to determine whether available data indicate that some compound or exposure should be considered a possible "hazard," and epidemiologists are needed to evaluate the strength of human studies, especially in attempts to determine whether an association between exposure and an adverse response is one of cause and effect. Dose-response modeling requires the input of statisticians, epidemiologists, and people expert in developing models that predict adverse response as a function of dose. Pathologists provide additional background on the nature of the adverse response, toxicologists are especially important for understanding mechanisms of toxicity and the relevance of animal data for human exposures, and bacteriologists may be critical in elucidating the spread of an infectious disease. Exposure assessment often requires the input of engineers, as well as hydrologists (for waterborne hazards), meteorologists (for airborne hazards), and analytical chemists. Industrial hygienists can be critical in providing insight into current and past occupational and environmental exposures; this insight may also be relevant to levels of exposure to the general population. The characterization of risk may involve all of these disciplines and many others.

Given the broad array of possible hazards of modern life and the complex issues raised by their assessment and possible control, it is no surprise that risk assessment, especially of chemical hazards, is difficult, plagued by uncertainty and often controversy. A major risk assessment (of lead, for example, or dioxin) can cost millions of dollars and require vast amounts of scientific talent. The need for reliable data of many kinds is enormous. Versions of the criteria of Hill[3] for inferring causality in epidemiology are important in this regard. These criteria include an appropriate temporal pattern, with exposure preceding response; a relation between increasing dose and increasing response; and the detection of the response across multiple studies conducted in different ways and in different populations. However, these criteria are not always appropriate. Epidemiologic studies may show an increase in the frequency of a congenital abnormality as exposures rise but a decrease in frequency at even higher doses because affected fetuses have such severe problems that most die in utero and cannot display the abnormality.

The risks associated with exposure to a hazard may be expressed by a variety of summary statistics that include individual lifetime risk, annual population risk, the percentage or proportion of increase in risk, and loss of life expectancy.[4] For example, individual risk might be an important feature of cigarette smoking, whereas years of potential life lost might be a relevant means of characterizing some occupational hazard, such as injuries, for which younger workers may be at special risk of unintended serious or fatal mishap.

Six essential issues arise in risk assessment. First, not every person exposed to a potential hazard will exhibit an adverse response. In addition, almost every adverse response to some exposure may occur even without exposure, although the link between asbestos and mesothelioma may be a near-exception. Thus, many long-term cigarette smokers escape without lung cancer, and nonsmokers sometimes get the disease, although people who smoke are still 10 to 15 times more likely than nonsmokers to get lung cancer.

Second, the frequency or magnitude of an adverse response generally depends on the degree and extent of exposure to a hazard, possibly with a threshold below which no risk is apparent. Many toxic drug reactions fit this pattern.

Third, people vary in their responses to the same level of dose or exposure. The risk for any individual may depend on a variety of intrinsic factors such as age, sex, prior or concurrent exposures to other hazards, and the level of detoxifying enzymes. Certain subgroups such as infants, the very old, and those with impaired immune systems may be at unusually high risk; this is often true for infectious diseases, but reasons for special sensitivity are often unknown.

Fourth, data for the direct measurement of human risk are often absent or seriously inadequate. Thus, the carcinogenic potential of a modest intake of saccharin, for example, is still uncertain because the primary evidence of carcinogenicity was from animal studies, in which doses were very high; the mechanism of carcinogenesis may not operate at low levels of exposure in humans.[5] (Similar issues arose in Canada with respect to cyclamate.) Conversely, biological processes can sometimes cause low-level exposures to be almost as hazardous as high-level exposures.[6]

Fifth, many risks are deemed acceptable, and their acceptability depends on many, sometimes surprising, factors, including the number of people exposed, whether exposure is voluntary, the social value of the exposure, mechanisms of compensation for harm or death, and familiarity with the risk.[7] We accept shockingly high rates of carnage on the highways because we value the freedom and convenience of personal transportation.

Finally, criteria are often unclear about the best way to balance risks and benefits to establish acceptable exposure limits for a hazard. For example, tamoxifen is clearly an ovarian carcinogen, but physicians continue to use this drug because of its great benefits as a chemotherapeutic and chemoprophylactic agent for breast cancer.

Risk in Context

We cannot avoid making decisions about risk management. To ignore them is to make those decisions by default, covertly and without full appreciation of their implications. Good risk management requires good risk assessment. The list of exposures that may be considered for risk assessment is quite long. Examples include a broad range of environmental and industrial chemicals (benzene, dioxin, asbestos), physical hazards (automobile accidents, noise, medical radiation), and biological agents (salmonella in hamburger, adverse reactions to vaccines). These and many other potential hazards are also closely linked to benefits we may want to retain (e.g., automotive transportation; effective and inexpensive industrial processes; vaccines). When benefits are important and perfect safety is unattainable, the acceptability of risk must be weighed with some care. Thus, estimation of the expected frequency and magnitude of outcomes under specific conditions of exposure and context, present or future, is critical when rational personal and societal choices may lead to some adverse outcomes.[1,2] Comparing, weighing, and choosing among different interventions to respond to risks is common in medical practice. Informing a patient who is about to have a laparoscopic appendectomy that there is a 2 percent chance that he or she will need an immediate laparotomy is an important part of the background for informed consent. One difference between medical practice and chemical risk assessment is that the 2 percent risk just mentioned may be based on hard data and much experience, whereas estimates of the risks of chemical exposures are often more speculative.

Two strategies are commonly used in quantitative risk assessment. One is the "margin-of-safety" approach, in which a scientific team looks for the highest dose that has produced no effect in animal or human studies, defined as the "no observed effect level" (NOEL), or sometimes for the lowest dose that did produce an adverse effect (LOEL). A set of "safety factors" is then applied, such as tenfold for using animal data rather than human data, another tenfold for the possibility that unexpected harm will arise later or in ways that have not been assessed, and still another tenfold when safety is especially important. These three factors multiplied together would lead to an exposure limit of 1/1,000th of the highest dose not known to cause problems in animals. The probability or size of risk at that point is not evaluated but is generally assumed to be virtually zero. A problem with this approach is that increases in knowledge about harmful effects will drive NOEL (and allowable exposures) downward, while that same greater knowledge may show that the safe limit of exposure is actually higher than had been assumed.

Margins of safety based on NOELs and safety factors have been widely used to evaluate systemic toxins and physical hazards, but quantitative regression modeling tools are more commonly used for carcinogens, and they are now the norm in risk assessments of many other adverse responses. These regression-based approaches fit a mathematical model to the data and are used to estimate the dose associated with a specified level of response. For example, in an animal tumorigenicity experiment, the proportions of animals with liver cancer at several doses (including zero dose, the control) may be used to estimate the risk added to background incidence by those and intermediate exposures. Interest often centers on exposures close to zero and far below any of the exposures in the animal study. As a result, scientists must make major assumptions—that is, educated guesses—about biological responses to low-dose exposures.

Uncertainty in Risk Assessment

The relation of risk to dose or exposure is generally unknown and often controversial. Should linear terms be required in the model? Should threshold parameters be included? Should one use point estimates of regression-based potency endpoints or lower confidence limits, which incorporate sampling variability? Although these questions may seem esoteric, different answers often lead to dramatically different conclusions about the size and nature of the risk. This is especially true in the prediction of responses to very-low-level exposures, the so-called low-dose extrapolation problem.

Unfortunately, accurate estimation of most risks is not possible. One example is the carcinogenicity of saccharin in the human diet. Very high doses of saccharin cause bladder cancer in animals; however, the biological mechanisms may have little relevance for humans, and data from human studies are limited, imprecise, and uncertain because almost all saccharin users have also used other artificial sweeteners that may have their own adverse effects. Another difficult situation occurs when estimating risks in humans, for example, who have been exposed intermittently to much lower levels of a chemical that has been found to be carcinogenic at high doses over a lifetime in small rodents, and these people have been followed up for a relatively short time. This is particularly vexing in the case of carcinogenic responses, where many years must elapse before cancer is detectable.[8] Finally, exposure to other agents may potentiate effects associated with the hazard of concern. The extra increase in risk of oral cancer in smokers who consume alcohol is an example of such synergy.[9] The uncertainties in risk assessment are so great that independent and technically competent reports on the same hazard often differ by a factor of a thousand or more.[10]

Using Risk Assessments

If risk assessments are so difficult and expensive but are still subject to great uncertainty, why should we bother with them? Arguably, what matters most is not the number(s) that one has at the end of the assessment, but the process that gets us there, although the public will ultimately want to know what level of exposure is "safe." A comprehensive risk assessment reviews, evaluates, and integrates the entire relevant literature to weigh the evidence for or against a hazard. A wide range of expertise will be brought to bear, and issues needing further study will be identified and refined. The needs of risk assessment provide powerful incentives for more and better science, including toxicology, pathology, epidemiology, biostatistics, industrial hygiene, exposure assessment, environmental transport studies (e.g., in air or groundwater), and many other fields of study. Risk assessment can always improve the foundation for risk management, and sometimes the best estimate of risk is so high or so low that risk-management decisions are clear despite the uncertainty.[10]

Differences among independent risk assessments based on the same information may engender a perception that the scientific community is clueless about the true risk associated with particular hazards. This problem is often complicated by a focus of media attention on the latest in a series of studies rather than the import of the whole of the evidence—an approach that leads to the "carcinogen of the week" mentality. When there are reports today that some exposure is unexpectedly hazardous, while tomorrow's study finds no such effect, the whole process may be considered shady and the credibility of science more generally damaged.

What is a prudent policy for triggering some response to a potential hazard? Some people require high levels of evidence before they will want to act (wait to see if there are bodies to count), while others require much lower levels (if a thing is at all doubtful, treat it as dangerous—the "precautionary principle"). Neither extreme seems to us to be appropriate, in part because risk management must integrate health risk assessment with such other things as economic analysis, legal mandates and constraints, level of public concern, and the availability of substitutes for useful but possibly risky products. In practice, the level of risk that triggers an intervention may differ between different exposure settings. For example, a one-in-a-million excess risk of some adverse response may be deemed acceptable for environmental exposures of the entire population, whereas a risk of one in a thousand may be deemed acceptable for some occupational exposures (we accept more than that for jockeys and steeplejacks), and one in ten may be acceptable for a medical intervention in desperate circumstances. Society accepts many risk-benefit trade-offs, such as those implicit in driving or flying. In essence, exposure levels are set at levels deemed to provide an acceptable trade-off between societal benefits and risks.[11]

Risk assessment reaches deep into our lives. Many members of the public may not want to learn all they would need to know to make reasoned choices about every possibly risky exposure, although some risks are well characterized and widely understood (e.g., cigarette smoking). Thus, there will be a continuing need for a cadre of strong scientist risk assessors, generally at the level of national government. In the United States, federal agencies with such interests include the Environmental Protection Agency, the National Institute of Environmental Health Sciences, the Food and Drug Administration, the Consumer Product Safety Commission, the Occupational Safety and Health Administration, and the

National Institute for Occupational Safety and Health. Similar agencies exist in most developed countries. An educated citizenry should understand that there is reason for centrally recommended exposure limits, that risk is not the only important thing in the regulation of hazards, that risk assessment is inherently uncertain (even when human data are available), and that our understanding of specific risks is likely to change only slowly with the accumulation of sound research studies, each adding a bit to the evidence for or against the existence of a hazard and our ability to quantify it accurately.

References

1. Committee on the Institutional Means for Assessment of Risks to Public Health, National Research Council. *Risk assessment in the federal government: Managing the process*. Washington, DC: National Academy Press, 1983.

2. Committee on Risk Assessment of Hazardous Air Pollutants, National Research Council. *Science and judgment in risk assessment*. Washington, DC: National Academy Press, 1994.

3. Hill AB. The environment and disease: Association or causation? *Proc R Soc Med* 1965; 58:295–300.

4. Cohrssen JJ, Covello VT. *Risk analysis: A guide to principles and methods for analyzing health and environmental risks*. Springfield, VA: National Technical Information Service, 1989.

5. Ellwein L, Cohen S. The health risks of saccharin revisited. *Crit Rev Toxicol* 1990; 20:311–326.

6. Bailar JC, Crouch EDC, Shaikh R, Spiegelman D. One-hit models of carcinogenesis: Conservative or not? *Risk Anal* 1988; 8:485–497.

7. Slovic P. *Perception of risk*. Science 1987; 236:280–285.

8. Bailar JC III. Current knowledge about the cancer latent period: The contribution of epidemiology. In: Maltoni C, Soffritti M, Davis W, editors. The scientific bases of cancer chemoprevention: *Proceedings of the International Forum on the Scientific Bases of Cancer Chemoprevention, 1996 March 31–April 2, Bologna, Italy*. Amsterdam: Elsevier, 1996. pp. 65–71.

9. Choi SY, Kahyo H. Effect of cigarette smoking and alcohol consumption in the aetiology of cancer of the oral cavity, pharynx and larynx. *Int J Epidemiol* 1991; 20:878–885.

10. Bailar JC, McGinnis JM, Needleman J, Berney B. *Methodologic challenges in health risk assessment*. Westport, CT: Auburn House, 1993.

11. Lowrance WW. *Of acceptable risk: Science and the determination of safety*. Los Altos, CA: Kaufman, 1976.

The Precautionary Principle: A Guide for Protecting Public Health and the Environment

Ted Schettler, Katherine Barrett, and Carolyn Raffensperger

14

Human activity has transformed the earth's ecosystems in unprecedented ways. These activities have fundamentally changed the surface of the earth and its atmosphere, including the cycling of carbon, nitrogen, water, metals, and the number and distribution of plants and animals.[1] Many environmental impacts of human activities cannot be predicted because of the complex interactions among variables in the ecosystem, which may lead to unanticipated global consequences.[2] In response to the capacity of humans to cause serious and often widespread harm to health and the environment, the precautionary principle was formulated as a guide to public policy decision making.[3]

The evidence of detrimental impacts is overwhelming.[4–6] Air, soil, and water quality are severely degraded throughout the world. The carbon dioxide concentration in the atmosphere has increased by nearly 30 percent in the last 150 years. Human activities are responsible for more atmospheric nitrogen fixation than all other sources combined, as well as for more mercury deposition on the surface of the earth than from other geological sources. Nitrates contaminate ground and surface water. Nitrous oxides are found in the air at toxic concentrations.

Freshwater and marine fish are sufficiently contaminated with mercury to require warnings to women of reproductive age to limit consumption because of risks to fetal brain development. Large numbers of plant and animal species have been driven to extinction, and most marine fisheries are severely depleted. Novel synthetic industrial chemicals contaminate the breast milk, egg yolk, ovarian follicles, and amniotic fluid of humans and other animals. The toxicity of most of these chemicals is not well known.

The precautionary principle states that with evidence of threats of significant harm, even in the face of scientific uncertainty, precautionary action should be taken to protect public health and the environment.[2] Recognizing that science and knowledge are always evolving and have limits, the precautionary principle addresses the need for a more refined consideration of threats of harm, scientific uncertainty, and actions taken. It is meant to enable and encourage precautionary actions that serve underlying values and that are based on what is known, while also taking into account what is not known. It encourages close scrutiny of all aspects of science—including the research agenda, funding, design, interpretation, and limits of studies—for potential impacts on the earth and its inhabitants.

The International Joint Commission (IJC) endorsed the precautionary principle in its *Sixth Biennial Report on Great Lakes Water Quality,* which addressed the manufacture and use of persistent, bioaccumulative toxic substances in the Great Lakes basin. The IJC found it disingenuous that science must prove with 100 percent certainty that an exposure will result in an adverse health effect before taking precautionary action. It concluded that "the focus must be on preventing the generation of persistent toxic substances in the first place, rather than trying to control their use, release, and disposal after they are produced."[8]

History and Definition of the Precautionary Principle

The term *precautionary principle* comes from the German *Vorsorgeprinzip*—literally, "forecaring principle." Its origins can be traced to German clean-air environmental policies of the 1970s that called for *Vorsorge,* or prior care, foresight, and forward planning to prevent harmful effects of pollution.[9] The precautionary principle has since been invoked in numerous international declarations, treaties, and conventions, and has been incorporated into the national environmental policies of several countries (see table 14.1). It has been applied to decisions on food safety, protection of freshwater systems, land development proposals, fisheries management, and the release of genetically modified organisms, among others.

One formulation of the precautionary principle is found in the 1992 Rio Declaration of nations participating in the United Nations Environ-

Table 14.1 Examples of treaties and conventions that explicitly invoke the precautionary principle

Montreal Protocol on Substances That Deplete the Ozone Layer (1987)
Ministerial Declaration of the Second World Climate Conference (1990)
Bergen Ministerial Declaration on Sustainable Development (1990)
Bamako Convention on Hazardous Wastes within Africa (1991)
Framework Convention on Climate Change (1992)
United Nations Conference on Environment and Development (1992)
Helsinki Convention on the Protection and Use of Transboundary Watercourses and International Lakes (1992)
Maastrict Treaty on the European Union (1994)
U.S. President's Council on Sustainable Development (1996)
Cartagena Protocol on Biosafety (2000)
Stockholm Convention on Persistent Organic Pollutants (2001)

ment Program treaty negotiations: "Where there are threats of serious or irreversible damage, lack of full scientific certainty shall not be used as a reason for postponing cost-effective measures to prevent environmental degradation."[10,11]

In 1998, a group of scientists, grassroots environmentalists, government researchers, and labor representatives from the United States, Canada, and Europe convened at the Wingspread Conference in Wisconsin to discuss ways to formalize and implement the precautionary principle. They formulated the precautionary principle as follows: "When an activity raises threats of harm to human health or the environment, precautionary measures should be taken even if some cause and effect relationships are not fully established scientifically" (Wingspread Statement, 1998).[3]

Simply put, the precautionary principle says that we should aim to anticipate and avoid damages before they occur or detect them early. It is based on underlying values and on the following core elements:

- Potential harm—predicting and avoiding harm, or identifying it early, should be a primary concern when contemplating an action
- Scientific uncertainty—the kind and degree of scientific uncertainty surrounding a proposed activity should be explicitly addressed
- Precautionary action—particular activity undertaken to avoid harm, even when the harm is not fully understood.

Elements of the Precautionary Principle

Underlying Values

Virtually every guide to public policy decision making that affects ecosystems and their inhabitants is based on an underlying set of values. The precautionary principle has its own values. The principle is based on recognizing that some activities may cause serious, irreparable, or widespread harm. Not all choices are "safe to fail," and the "endless" frontiers to which our ancestors could move to begin anew generally no longer exist.

The principle is also based on the assumption that people have a responsibility to preserve the natural foundations of life, now and into the future. The origin of that responsibility differs among people who share it. Using an anthropocentric utilitarian analysis, some people recognize that ecosystems provide essential services, such as clean air, water, and food necessary for human health, and these should be protected for current and future generations. Other people consider both humans and nonhuman species to have the right to life and other protections, such as freedom from cruel treatment or gratuitous exploitation. Others believe that the proper focus of ethical consideration is the integrity of entire ecosystems rather than individual species. Some are motivated by the notion of stewardship, which requires taking care of all species to the extent possible. Spiritual considerations, beauty, and sympathy are critical for many. The precautionary principle does not require a resolution of these differences. Decision makers, however, should examine the basis for their decisions through the ethical lens of taking care and precaution.

Not everyone accepts responsibilities to global ecosystems or future generations of any species. Many past and current public policies fail to consider seriously the future quality and integrity of life-sustaining ecosystems. Antithetical to the precautionary principle is human activity that treats the world as a collection of resources to be extracted, consumed, and discarded, with little understanding of ecological principles and little regard for long-term impacts. Although it is not possible to predict the needs of future generations or the future condition of ecosystems, it is certain that what is done now will have future consequences.

A precautionary approach is based on determining how much harm can be avoided rather than deciding how much harm is acceptable or how much can be assimilated. It acknowledges that the world consists of complex, interrelated systems, vulnerable to harm from human activities and resistant to full understanding. Precaution is an expression of the value of giving priority to the protection of these vulnerable systems. This prescriptive aspect of the principle has been expressed both as general duty to avoid harm and as a right of communities and governments to adopt precautionary measures. For example, it is expressed as a duty in the Rio Declaration and Wingspread Statement and as a right in the Cartegena Protocol on Biosafety.

The Potential for Harm

Precautionary action is appropriate when there is some evidence that a particular technology or activity might be harmful, even if the nature of that harm is not fully understood. This means that decision makers must consider potential hazards that have been, to some extent, identified or that are plausible, based on experience, what is known, and/or what can be predicted. Threats of serious, irreversible, cumulative, or widespread harm are obviously of greater concern than trivial threats.

Harm can occur at the level of the cell, organism, population, or ecosystem. Impacts may be biological, ecological, social, economic, cultural, or political. They may be distributed equally or disproportionately among individuals or populations or geographically, now or in the future. Because systems are complex and outcomes are not always predictable, decision makers must identify parameters to assess the potential impacts of a proposed activity. Moreover, the standard against which an impact is measured must also be defined. Asking whether a proposed pesticide is more environmentally friendly than another, for example, is a very different question from asking whether either is necessary at all. Cities in the Northeast United States facing newly emerging threats from mosquito-borne West Nile viral encephalitis are currently debating this very issue.

Scientific Uncertainty

Recognizing the uncertainty and limits of science is central to the precautionary principle. It is not always possible to predict the consequences of an action in complex social systems or ecosystems.[12] When inputs are modified, the behavior of complex systems is often surprising. For example, the widespread industrial use of chlorofluorocarbons (CFCs) directly led to stratospheric ozone depletion and a consequent increase in ultraviolet light exposure and skin cancer risk. By the time impacts are documented, considerable harm may have occurred. Despite early warnings, the use of lead in gasoline and paint, for instance, damaged the brain function of generations of children.[13] It is also possible for a system to cross a threshold and operate at a new state of relative equilibrium, from which there is no turning back. Global trade and travel may, for example, introduce bacteria, viruses, insects, and other exotic species into ecosystems where they did not previously exist, allowing them to become established and cause harm.

Understanding cause-and-effect relationships in complex systems is limited by different kinds of uncertainties, some of which are more easily reduced than others.[12] Uncertainty often results from more than a simple lack of data or inadequate models. Some kinds of uncertainty simply cannot be reduced because of the nature of the problem being studied.

Most complex problems have a mixture of three general kinds of uncertainty—statistical, model, and fundamental—each of which should be considered before deciding how to act.

Statistical Uncertainty

Statistical uncertainty is the easiest to reduce or to quantify with some precision. It results from not knowing the value of a particular variable at a point in time or space but knowing, or being able to determine, the probability distribution of the variable. For instance, in the case of IQ distribution among a population of individuals, valid decisions can be made based on knowing the likelihood of a variable having a particular value.

Typically, however, real-world decisions are made in the context of multiple, interactive variables. For example, the incidence of cancer that is attributable to exposure to a carcinogen among a genetically and geographically diverse population will differ from that in a group of genetically similar rodents living in controlled laboratory conditions. Similarly, understanding the global climate impact of a proposed shift in energy policy requires more than simply knowing the probability distribution of a single variable. When more than one variable is involved, a model is typically constructed with certain assumptions and simplifications, introducing a new kind of uncertainty.

Model Uncertainty

Model uncertainty is inherent in systems consisting of multiple variables interacting in complex ways. Even if the statistical uncertainty surrounding the value of a single variable can be defined or reduced, the nature of relationships among system variables may remain difficult to understand. This is particularly problematic for any model of complex systems. We might conclude that there will be a tendency for the system to behave in a certain way, but the likelihood of that behavior is difficult to estimate.

Complex models can include only a finite number of variables and interactions. The real world, however, is a confluence of biological, geochemical, ecological, social, cultural, economic, and political systems. No experimental model can fully account for each of these and their interrelationships. Ongoing research, monitoring, and model refinement may help to reduce uncertainties, but imprecision and surprise are inevitable. Indeterminacy, which increases when moving from statistical to simple to complex model uncertainty, is, at some point, more correctly called ignorance.

Fundamental Uncertainty

Fundamental uncertainty encompasses this ignorance. Ignorance can either be internal ("I don't know, though someone does") or external ("no one knows"—a result of the complexity or uniqueness of the system). External ignorance is inherent in novel or complex systems where existing models do not apply.

This type of fundamental uncertainty can result from having no valid knowledge of the likelihood of a particular outcome. Consider, for example, attempts to understand the relative contributions of overfishing, habitat destruction, pollution, climate change, and eutrophication to declines in coastal marine fish stocks. Interactions among these variables are likely to change, depending on the state of the system at a given time. A historical analysis suggests that in early human history, fish stocks were so abundant that overfishing was the most important variable. But as fish populations declined, the entire coastal marine ecosystem was fundamentally altered so that pollution, eutrophication, and habitat destruction became increasingly significant.[14] Fisheries management and restoration efforts will always face some fundamental uncertainties about ecosystem conditions prior to the onset of overfishing and how interactions among variables will change under changing circumstances.

Fundamental uncertainty can also result from having no knowledge of what outcomes can occur. Here we do not even know what we do not know. Chemical regulators, for example, were unaware of the existence and functions of a stratospheric ozone layer that would be damaged by CFCs when allowing them to be marketed as safe for commercial use. Fundamental uncertainty is extremely difficult to reduce or otherwise manage and demands awareness, respect, and humility.

Scientific Uncertainty and Scientific Proof

These kinds of uncertainty must be kept in mind when considering the notion of scientific proof. Scientific proof is a value-laden concept that integrates statistics, empirical observation, inference, research design, the research agenda, and politics in a social context. Proof is the result or effect of evidence as well as the establishment of a fact by evidence.[15] By the time a fact or causal relationship has been established by rigorous standards of proof, however, considerable avoidable harm may already have occurred, as with the impacts of lead on brain development and asbestos on lung cancer risk.

By convention, a rigorous standard of evidence is necessary to establish factual "proof" of a cause-and-effect relationship. Traditionally, in a study of the relationship between two variables, a correlation is said to be established with statistical significance only if the results show the two to be linked, independent of other factors, with a greater than 95 percent likelihood that the results of the study truly depict the real world. But correlation does not establish causation. In epidemiology, a series of additional criteria—for example, those of Hill—are usually added before causation can be claimed.[16] In addition to establishing a statistically significant correlation between two variables, Hill criteria require that the causal variable precede the effect, a dose-response relationship be established, sources of bias and confounding be eliminated, coherence with other studies be achieved, and a plausible biological mechanism be proposed. Tobacco smoking, for instance, was long known to be associated with lung cancer, but when a plausible biological mechanism was finally described, it became impossible to deny that tobacco smoking "causes" cancer.

Strict criteria may be useful for establishing "facts," but they are unlikely to routinely protect public health and the environment, particularly when impacts are delayed or when potentially causal factors are difficult to study. For example, ethical considerations generally preclude studying the impacts of chemical exposures on humans under controlled experimental conditions (with the exception of pharmaceuticals). Instead, a series of significantly positive epidemiological studies, along with in vivo and in vitro laboratory test results, are often deemed necessary to establish "proof" of toxicity and harm. Meanwhile, lack of "proof" of harm is often used to justify continued human and ecosystem exposures to inadequately studied chemicals. Given the limits of scientific inquiry and of its practical application in complex systems, establishing a high bar of proof as a prerequisite for taking action has the potential to result in unnecessary and often irreversible harm.[17,18]

Under the precautionary principle, the burden of proof could shift from one party to another, depending on weight of evidence, lack of evidence, scientific uncertainty, and the nature of the harm of concern. Standards of evidence for demonstrating harm (or safety) could differ from those under policies that require conclusive proof before remedial or preventive action is warranted. For example, standards of evidence could be similar to those used in occupational health law or tort cases, the standard being "more likely than not." This is different from criminal cases, which require evidence that establishes guilt beyond a reasonable doubt.

When dealing with complex systems, evidence is rarely sufficient to quantify or predict the consequences of human activity beyond doubt. Yet the justification for any action can only be based on information that is available. Inaction, based on lack of quantifiable proof of harm, is, in itself, a form of action.

Precautionary Action

To serve the values that underlie the precautionary principle, action should be anticipatory in order to prevent harm to public health and the environment, despite underlying scientific uncertainty. The precautionary principle does not specify which actions are appropriate under particular circumstances. It is a guiding principle, not a set of binding rules. The choice(s) among potential anticipatory actions, however, should be informed by the following:

- Full consideration of the weight of evidence for potential harm
- The kind and degree of scientific uncertainty associated with that evidence
- An assessment of potential alternative actions

Implementing the Precautionary Principle

Policy decisions are often made by focusing primarily on economic or political factors or by considering only one kind of risk. The precautionary principle requires a systematic look at the potential for various kinds of harm, taking into consideration scientific uncertainty and underlying fundamental values before deciding what to do.

Goal Setting

Goal setting is particularly important for establishing environmental and health policies because choices among human activities can have widespread adverse or favorable impacts on regional and global ecosystems.[19] Goal setting requires us to ask the following questions: (1) Where do we want to be at some future time? (2) Who should have an opportunity to answer that question? (3) What mechanisms are in place or could be put in place to determine which visions for the future will prevail? Examining agreed-on goals for the future can be useful for developing a strategy for getting there. Of course, not all goals are generally agreed on or represent shared visions. But as goals are made explicit, the values and assumptions underlying decision-making processes will also become more transparent and may result in processes for reconciling differences.

Technologies designed to solve narrowly defined problems or goals are likely to be reconsidered if more far-reaching questions are asked. For example, monocropping trees in plantations may supply short-term profits to corporations and wood to homebuilders, but it is also likely to destroy forest habitat critical for numerous species and ecosystem functions over the long term.

Assessing Alternatives

A truly precautionary approach requires examining a range of options for meeting policy goals. Currently, there are few requirements for comprehensively assessing a range of alternatives to proposed activities. For instance, current regulatory policies emphasize a risk-assessment/risk-management framework. Risk assessment attempts to estimate the probability of harm (risk) from a proposed activity and then asks whether that harm is acceptable. Risk-management techniques are intended to minimize the risks of the proposed activity, but not to question whether the activity is necessary for achieving broader goals.

Alternatives assessment instead asks whether the harm is necessary and if there might be other ways to achieve agreed-on goals that would avoid harm altogether. When alternatives assessment is applied early in policy decision making, innovative approaches that reflect societal goals, ecological principles, and the values that underlie the precautionary principle are more likely to emerge. Assessing alternatives can also lead to actions that truly respect the level of uncertainty in given circumstances.

In many cases, the appropriate choice among alternatives may not be obvious because of lack of experience or limited ability to predict distant consequences. In these instances, certain ecological principles may be useful guides. Aldo Leopold, for example, encouraged examination of each question "in terms of what is ethically and esthetically right, as well as what is economically expedient. A thing is right when it tends to preserve the integrity, stability, and beauty of the biotic community. It is wrong when it tends otherwise."[20] The Natural Step, originating from the work of Karl-Henrik Robert of Sweden in the late 1980s, is based on four general principles.[21] They follow from the conclusion that sustainability depends on returning to the earth as reuseable waste whatever is removed for food, energy, or other forms of consumption. The Natural Step holds that (1) substances from the early's crust must not systematically increase on the surface of the earth, (2) substances produced by society must not systematically increase in the environment, (3) the physical basis for productivity and diversity in the natural world must not be systematically diminished, and (4) we must be fair and efficient in meeting basic human needs. When applied, these principles often lead to decisions that differ considerably from those based on short-term or otherwise limited considerations.

Adopting Transparent, Inclusive, and Open Processes

Many decisions likely to have significant impacts on ecosystem health have historically been made behind closed corporate and government doors. Public policy is driven by narrowly construed risk assessments and cost-benefit analyses, influenced by simplifications and assumptions that are rarely explicit and that often fail to reflect inherent uncertainties.

When the health and well-being of the public and environment are at stake, a precautionary approach requires open, inclusive, and transparent processes that are initiated early in decision making, beginning with goal setting. This participatory approach is justified not only by the fundamental fairness of democratic decision making but also because a broad range of knowledge and experience leads to better science, more comprehensive understanding of the potential for harm, and better decisions in keeping with societal goals. Transparent processes also ensure that the basis for actions is substantiated and that decision makers are accountable.

Analyzing Uncertainty

Virtually every assessment of the impacts of a proposed activity or policy decision is limited by some kind or degree of uncertainty. A precautionary approach requires explicit recognition of the nature of uncertainty that is inherent in understanding the potential for harm from an ongoing or proposed activity.

Statistical uncertainty may be reduced with more data collection. Model and fundamental uncertainty, however, are not as easily handled. When the latter predominate, a requirement to resolve uncertainty as a prerequisite for decision making shows a fundamental lack of understanding of the limits of science, or it may be nothing more than a tactic to maintain the status quo. President George W. Bush, for example, used scientific uncertainty about the impacts of greenhouse gases on global climate to justify U.S. rejection of the Kyoto treaty on global climate change and to promote an energy policy weighted heavily in favor of increasing fossil-fuel extraction and consumption, with virtually no emphasis on conservation or renewable sources.

A precautionary approach relies on a weight-of-evidence analysis, in which evidence is gathered from numerous, diverse sources and considered in a deliberative manner. Scientific studies are often based on hypothesis testing. Typically, a null hypothesis is formulated stating, for example, that an activity to be evaluated is *not* associated with a harmful outcome. The data gathered in the study either support the null (no association with the outcome of concern) or refute it (an association is observed). It is customary to require that the data strongly refute the null hypothesis, with a high degree of significance, in order to conclude that an association really does exist. This requirement biases data interpretation toward making a Type II error (false negative), deciding that a proposed

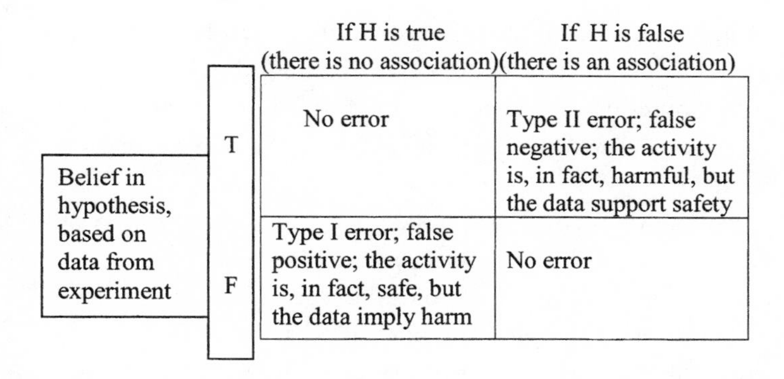

Figure 14.1 Null hypothesis (H): No association exists between activity and outcome of concern

activity is likely to be safe when it is, in fact, harmful, rather than a Type I error (false positive), deciding that a safe activity is likely to be harmful (see figure 14.1).

When serious, irreversible harm is plausible from a proposed activity, shifting interpretive biases, by decreasing the likelihood of Type II errors (false negatives), facilitates protection of public health and the environment. As a result, however, the possibility of Type I errors (false positives) increases. Each kind of error has consequences, and each must be considered in context.

Burden of Proof and Responsibility

Under the precautionary principle, the burden of proof regarding the safety of an activity may shift. Certain products and industries in the United States are subject to regulatory oversight. Manufacturers of some products, such as pesticides, are required to supply some basic prospective safety-evaluation information in order to receive permission to market the product. For most products and activities, however, the onus falls on the general public or on the govenrment to demonstrate that harm is likely. For example, with rare exceptions, manufacturers of nonpesticidal industrial chemicals are not required to submit comprehensive safety or toxicity testing data, other than what the manufacturer may have chosen to gather, before announcing their intention to bring a product to market. Instead, the administrator of the U.S. Environmental Protection Agency must make the case within 90 days that these data are necessary.[22] When uncertainty is significant and difficult to resolve, and when evidentiary requirements are set at a high level, proof of the likelihood of harm is difficult to establish. Consequently, an activity may continue to be permitted that could otherwise be modified or stopped in favor of a less harmful alternative.

For the purposes of establishing public policy, the precautionary approach suggests that the burden of proof is better thought of as the burden of persuasion and responsibility. This avoids the fruitless assertion that absolute safety can never be "proven." It acknowledges that, as the potential for serious, irreversible harm and scientific uncertainty increase, the proponent of an activity has an increasing obligation to account for the safety of the activity and to take responsibility for adverse impacts that may result from it. The onus for comprehensive testing, monitoring, and assumption of liability shifts to the proponent.

Learning and Adaptation

The precautionary approach incorporates processes that maintain flexibility and allow for learning and adaptation. Appropriate research and monitoring are essential. Decisions may be systematically revisited, based on new information. Efforts to restore ecosystems and avoid continued damage can be promoted. Through labeling or prenotification, potentially affected people are kept informed about products or activities that may directly affect them. The research agenda of private and public institutions may be designed to reflect broad social goals that extend well beyond developing marketable products. In this way, the precautionary approach is not linear but iterative, with feedback loops that search for and take into account new information and unintended consequences of provisional decisions.

Options for Precautionary Action

Precautionary action can be carried out in a number of ways. At the level of regulation, when research and development of a product or technology are complete, and only regulatory approval is needed for production and marketing, the options are ordinarily limited to yes, no, with limits, with monitoring, with labeling, or with posting of a performance bond. Commonly, a full assessment of need and alternatives is not permitted at the regulatory level.

The precautionary principle is better applied at a preregulatory level, when it can be used to encourage a closer look at problems that proposed technologies are intended to solve. How was the problem defined and by whom? Was the problem framed in the only or best way? Are there alternatives to the proposed technology?

The Relationship of the Precautionary Principle to Risk Assessment

A risk-assessment/risk-management approach to public policy decision making predominates in the United States. With few exceptions, risk assessments attempt to estimate the potential risks of proposed products or activities on a case-by-case basis without consideration of the complete context in which the activity will be carried out and rarely with any consideration of alternatives to the proposal.[23] Risk assessment supports a fragmented view of the world that promotes business as usual.

The precautionary principle presents a different view based on a number of factors, including how much we know, how much we can know, how broadly questions should be framed, which questions should be asked, who should frame the questions, the value of nonhuman life, our responsibility to future generations, and how we plan for the future.

Quantitative risk assessments usually respond to narrowly framed questions, and they are often limited by simplifications and assumptions. Risk assessments almost always fail to consider a full range of biological, ecological, social, cultural, and economic impacts and how they are distributed among individuals, populations, or communities. Advocates of a regulatory system dominated by quantitative risk assessments argue that they are inherently precautionary because they use conservative assumptions and safety factors. Risk assessors, however, often fail to distinguish among various kinds of uncertainty and tend to misclassify some model and fundamental uncertainty as statistical, to which they apply "uncertainty" factors. When model and fundamental uncertainty predominate in a system, this approach may lead to large underestimates of risk or uncertainty as to the probability distribution of risk. This approach may also fail to predict adverse impacts removed in time and space, and will completely fail to predict surprises or novel impacts. Model and fundamental uncertainty predominate, for example, in understanding global climate change, the impacts of loss of biodiversity, exposures to real-world mixtures of industrial chemicals, and environmental causes of cancer.

Risk assessors often claim that the precautionary principle is "anti-science" or a tool to keep certain technologies from the marketplace. In fact, a precautionary approach encourages more science rather than less, acknowledging the need for precautionary action while addressing scientific uncertainty that may be intractable using available tools.

Decision making in the face of uncertainty is, of course, necessary, is frequently difficult, and requires assessment of relative risks. Guided by an overarching precautionary principle, however, these assessments are not the exclusive domain of risk analysts. They can be fully participatory and include a full consideration of a range of alternatives.

Conclusion

The precautionary approach is based on the ethical notions of taking care and preserving the natural foundations of life. It encourages research, innovation, and cross-disciplinary problem solving. If current trends of a number of environmental quality indicators continue, life as we know it for humans and many other species is not sustainable. We can anticipate that many more species will be driven to extinction and more people will live seriously impoverished and chaotic lives in an increasingly unstable and artificial world.[24] The struggle over who will have the means to remain relatively unaffected will become increasingly complex and difficult.

The precautionary principle helps to address these circumstances. It brings together ethics and science, illuminating their strengths, weaknesses, values, or biases. The precautionary principle serves as a guide for considering the impacts of human activities and provides a framework for protecting life-sustaining systems for future generations.

Notes

1. Vitousek P, Mooney H, Lubchenco J, Melillo J. Human domination of earth's ecosystems. *Science* 1997; 277: 494–499.

2. Myers N. Environmental unknowns. *Science* 1995; 269: 368–369.

3. Raffensperger C, Tickner J, editors. *Protecting public health and the environment: Implementing the precautionary principle.* Washington DC: Island Press, 1999.

4. Lubchenco J. Entering the century of the environment: A new social contract for science. *Science* 1998; 279: 491–497.

5. US Environmental Protection Agency. *Mercury study report to Congress.* EPA-452/R-97-003. Washington, DC: US Environmental Protection Agency, 1997.

6. Johnson N, Revenga C, Escheverria J. Managing water for people and nature. *Science* 2001; 292: 1071–1072.

7. International Joint Commission. *Sixth biennial report on Great Lakes water quality.* Windsor, ON: International Joint Commission, 1992.

8. International Joint Commission. *Sixth biennial report on Great Lakes water quality.* Windsor, ON: 1992: http://www.ijc.org/comm/6bre.html chap 2.

9. Boehmer-Christiansen S. *The precautionary principle in Germany—enabling government.* In: O'Riordan T, Cameron J, editors. Interpreting the precautionary principle. London: Earthscan, 1994.

10. The "Rio Declaration" on environment and development. The United Nations Conference on Environment and Development, June 1992, Principle 15.

11. Shabecoff P. *A new name for peace: International environmentalism, sustainable development, and democracy.* Hanover, NH: University of New England Press, 1996. pp. 86, 156, 172.

12. Perrow C. *Normal accidents: Living with high risk technologies.* New York: Basic Books, 1984.

13. Markowitz G, Rosner D. "Cater to the children": The role of the lead industry in a public health tragedy, 1900–1955. *Amer J Pub Health* 2000; 90(1): 36–46.

14. Jackson J, Kirby M, Berger W, Bjorndal K, Botsford LW, Bourque BJ, et al. Historical overfishing and the recent collapse of coastal ecosystems. *Science* 2001; 293(5530): 629–638.

15. For definitions, see *Black's Law Dictionary.* Garner BA, ed. 7th ed. Eagan, MN: West Group, 1999.

16. Hill AB. The environment and disease: Association or causation? *Proc R Soc Med* 1965; 58:295.

17. Beauchamp DE, Steinbock B, editors. *New ethics for public's health.* New York: Oxford University Press, 1999.

18. Kriebel D, Tickner J, Epstein P, Lemons J, Levins R, Loechler EL, et al. The precautionary principle in environmental science. *Environ Health Perspect* 2001; 109(9):871–876.

19. Cairns J. Defining goals and conditions for a sustainable world. *Environ Health perspect* 1997; 105: 1164–1170.

20. Leopold A. *A Sand County almanac.* New York: Ballantine Books, 1970, p. 262.

21. The Natural Step: Where science and business meet sustainability. *Business and the Environment* 1996; 7(7). Other details available at www.naturalstep.com

22. Toxic Substances Control Act. Public Law 94-469. Sect. 4. 1976.

23. O'Brien M. *Making better environmental decisions: An alternative to risk assessment.* Cambridge, MA: MIT Press, 2000.

24. National Research Council. *Our common journey: A transition toward sustainability.* Washington, DC: National Academy Press, 1999.

Vulnerable Populations

Philip Landrigan and Anjali Garg

15

Children and workers are two groups within the population who are particularly vulnerable to toxic hazards in the environment. The vulnerability of infants and children reflects their unique patterns of exposure to environmental hazards, coupled with the inherent fragility of their developmental processes. Workers' vulnerability is a consequence of the fact that their exposures to toxins are often magnitudes higher than those of the general population. In addition, workers may be exposed to newly developed and inadequately tested toxic materials and hazardous processes months or years earlier than the general public; thus, new diseases of toxic origin are often first recognized in this group.

In traditional risk assessment and regulatory practice, the special vulnerabilities of infants, workers, and other vulnerable subgroups (e.g., elderly or immune-compromised persons) have generally been ignored. Traditional risk assessment and regulation for community populations have long operated under the outmoded fiction that the entire population can be assumed to consist of adult white males weighing 70 kilograms, and standards have been set at levels sufficient to protect the health and safety of such a hypothetical population.

In this chapter, we explore the concept that certain groups within the population, most notably children and workers, are deserving of special protection in risk assessment, law, and regulation. We review data on the special vulnerabilities of these groups and explore the implications of those findings for public health.

Children

Patterns of illness have changed dramatically among children in the United States over the past century.[1] The prevalence of the classic infectious diseases has been greatly reduced, and many of these diseases have been brought under control.[2] Infant mortality has been lowered, although not equally across American society, and life expectancy at birth has increased by more than 20 years. Today, the most serious diseases confronting children in the United States are a group of chronic, disabling, and sometimes life-threatening conditions termed the "new pediatric morbidity." Examples include asthma, for which mortality has more than doubled between 1982 and 1992;[3] childhood cancer, for which incidence of certain types has increased substantially;[4,5] neurodevelopmental and behavioral disorders;[6,7] diseases caused by environmental tobacco smoke;[8–10] and certain congenital defects of the reproductive organs such as hypospadias, for which incidence in the past two decades has doubled.[11]

The causes of the chronic diseases and developmental disabilities that are so prevalent today among America's children are not well understood. An estimated 10 to 20 percent of these diseases appear to be of familial or genetic origin. Some are the consequence of obstetrical difficulties, infections, or trauma. But for the remainder, the causes are only beginning to be known.

In recent years, it has become increasingly recognized that children's exposures to toxic chemicals in the environment cause or contribute to the causation of certain of these diseases.[12,13] Lead is now recognized as a significant cause of neurobehavioral impairment, producing measurable disability (loss of IQ) at very low levels in blood, as well as alteration of behavior. Exposures in utero to polychlorinated biphenyls (PCBs) and methylmercury can also cause loss of intelligence.[14] Certain pesticides appear to interfere with brain development and reproductive function.

Children are extensively exposed to synthetic chemicals in the environment. More than 80,000 new synthetic chemical compounds have been developed and disseminated in the environment over the past 50 years. Children are at special risk of exposure to the 2,800 high-volume chemicals that are produced in quantities greater than one million pounds per year and that are most widely dispersed in air, water, food crops, communities, waste sites, and homes.[15] Fewer than half of these high-volume chemicals have been tested for their potential toxicity, and fewer still for their possible developmental toxicity to fetuses, infants, and children.[16,17]

Children's Unique Vulnerability

Children are highly vulnerable to chemical toxins.[18,19] They have disproportionately heavy exposures to environmental toxicants. Pound for pound of body weight, children drink more water, eat more food, and breathe more air than adults. Children in the first six months of life drink seven times as much water as the average American adult. One- to five-year-old children eat three to four times more food. The air intake of a resting infant is twice that of an adult. The implication of these findings for health is that children will have substantially heavier exposures than adults to any toxicants present in water, food, or air. Two additional characteristics of children further magnify their exposures: their hand-to-mouth behavior and their play close to the ground.

Children's metabolic pathways, especially in the first months after birth, are immature. Their ability to metabolize, detoxify, and excrete many toxicants differs from that of adults. In some instances, children are actually better able than adults to deal with environmental compounds, such as polyaromatic hydrocarbons and estrogen. More commonly, however, they are less well able to deal with chemical toxins such as lead and organophosphate pesticides because they do not have the enzymes necessary to metabolize them and thus are more vulnerable to them.[20]

Children undergo rapid growth and development, and their developmental processes are easily disrupted. Organ systems in infants and children undergo very rapid change prenatally, as well as in the first months and years after birth. These developing systems are very delicate and cannot easily repair damage that may be caused by environmental toxicants. Thus, if cells in an infant's brain are destroyed by chemicals such as lead, mercury, or solvents, or if false signals are sent to the developing reproductive organs, there is high risk that the resulting dysfunction will be permanent and irreversible.

Because children have more future years of life than most adults, they have more time to develop chronic diseases triggered by early exposures. Many diseases caused by toxicants in the environment require decades to develop. Many of those diseases, including cancer and neurodegenerative diseases, are now thought to arise through a series of stages that require years or even decades to evolve from earliest initiation to actual manifestation of disease. Carcinogenic and toxic exposures sustained early in life, including prenatal exposures, appear more likely to lead to disease than similar exposures encountered later.

Case Studies

Understanding of children's vulnerability to environmental toxins traces its origins to early studies of major disease outbreaks of toxic origin in children. These analyses formed the basis for our current understanding that children are uniquely sensitive to many toxins in the environment. The following paragraphs describe some of these reports.

A report from Queensland, Australia, in 1904 described an epidemic of lead poisoning in young children.[21] Clinical and epidemiologic investigation traced the source of the outbreak to the ingestion of lead-based paint by children playing on verandas. This was the first report of lead poisoning in children, and it led to the banning of lead-based paint in many nations, although not in the United States.

Studies of an epidemic of leukemia in the 1950s among young children in Hiroshima and Nagasaki documented their exposure to ionizing radiation in the atomic bombings.[22] These and subsequent studies of fetuses exposed in utero[23] established that infants and fetuses are more sensitive to leukemia following radiation than adults. Additionally, an increased risk of microcephaly was observed among infants exposed to radiation as fetuses in the first trimester of pregnancy during the atomic bombing.[24]

A report from Minamata, Japan, in the 1960s described an epidemic of cerebral palsy, mental retardation, and convulsions among children living in a fishing village on the Inland Sea.[25] This epidemic was traced to ingestion of fish and shellfish contaminated with methylmercury. The source of the mercury was found to be a plastics factory that had discharged metallic mercury into the sediments on the floor of Minamata Bay. The mercury was transformed by microorganisms into methylmercury, and it then bioaccumulated as it moved up the marine food chain, eventually reaching children who ate fish and shellfish. The most devastating effects were seen among children exposed in utero.

Subclinical Toxicity

A critically important intellectual step in the development of understanding of children's special vulnerability was the recognition that environmental toxins can exert a range of adverse effects in children. Some of these effects are clinically evident, but others can be discerned only through special testing and are not evident on the standard examination, hence the term *subclinical toxicity*. The underlying concept is that a dose-dependent continuum of toxic effects exists, in which clinically obvious effects have their subclinical counterparts.[26]

The concept of subclinical toxicity traces its origins to pioneering studies of lead toxicity in clinically asymptomatic children undertaken by Herbert Needleman and colleagues.[27] Needleman et al. showed that children's exposure to lead could cause decreases in intelligence and alteration of behavior even in the absence of clinically visible symptoms of lead toxicity. The subclinical toxicity of lead in children has subsequently been confirmed in prospective epidemiologic studies.[28] Similar subclinical neurotoxic effects have been documented in children exposed in utero to PCBs[29] and to methyl mercury.[30]

The Need for Prevention

The protection of children against environmental toxins is a major challenge to modern society. Hundreds of new chemicals are developed every year and are released into the environment. The majority of these chemicals are untested for their toxic effects on children.[15] The challenge is to design polices that specifically protect children against environmental toxins. To meet this challenge, a new paradigm for environmental health policy needs to be developed that is centered on the sensitivities and exposures of children. The analysis then begins with the child, his or her biology, exposure patterns, and developmental stage. The paradigm calls for a new way of thinking and a retooling of the risk-assessment process so that it takes into account the increased vulnerability of children.

The issue of environmental exposure looms increasingly large from a global standpoint. It is imperative that we develop policies that will protect the health of our children now and in the future.

Workers

Workers may be exposed to high concentrations of toxic agents and often are exposed earlier than members of the general population. Many environmentally induced diseases have first been seen in working populations because workers typically constitute well-defined groups, the nature and extent of their exposures are known, and they lend themselves to epidemiologic analysis. The appearance of illnesses in workers may provide a warning to the general population of the toxicity of environmental toxins.

At the same time, occupational disease is not limited to the workplace. Toxic hazards may be released from the workplace into the community environment, polluting the air, drinking water, or food chain.[31] Occupational toxins may also be transported home on the clothing of contaminated workers, causing such illness as lead poisoning and mesothelioma in family members.[32,33]

Principal Types of Occupational Disease

Occupational illness can affect virtually every organ system. It is beyond the scope of this chapter to review the full range of occupational illnesses; several textbooks on occupational medicine are available.[34–36] To develop a systematic approach to the prevention of illness of occupational origin and to rank categories of occupational disease according to relative priority, the National Institute for Occupational Safety and Health (NIOSH) has developed a list of the ten most important categories of occupational illness.[37,38] (See table 15.1.)

Diagnosis of Occupational Disease

Occupational diseases are underdiagnosed, and many diseases of occupational origin are incorrectly attributed to other causes. The underdiagnosis of occupational illness reflects that, clinically and pathologically, most occupational illnesses are not distinct from chronic diseases associated with nonoccupational etiologies. For example, lung cancer caused by asbestos is identical in its clinical and pathological presentation to lung cancer caused by cigarette smoking.[39] Similarly, solvent-induced encephalopathy may be attributed to old age.[40] Only in rare instances, such as in the association between asbestos and mesothelioma[39] or in that between vinyl chloride monomer and angiosarcoma of the liver,[41] is the causal asso-

Table 15.1 NIOSH list of ten leading work-related diseases and injuries.*

1. Occupational lung diseases:
asbestosis, byssinosis, silicosis, coal workers' pneumoconiosis, lung cancer, occupational asthma

2. Musculoskeletal injuries:
disorders of the back, trunk, upper extremity, neck, lower extremity; traumatically induced Raynaud's phenomenon

3. Occupational cancers (other than lung):
leukemia; mesothelioma; cancers of the bladder, nose, and liver

4. Severe occupational traumatic injuries:
amputation, fracture, eye loss, laceration, and traumatic death

5. Occupational cardiovascular diseases:
hypertension, coronary artery disease, acute myocardial infarction

6. Disorders of reproduction:
infertility, spontaneous abortion, teratogenesis

7. Neurotoxic disorders:
peripheral neuropathy, toxic encephalitis, psychoses, extreme personality changes (exposure-related)

8. Noise-induced loss of hearing

9. Dermatological conditions:
dermatoses, burns (scaldings), chemical burns, contusions (abrasions)

10. Psychological disorders:
neuroses, personality disorders, alcoholism, drug dependency

Source: Reference 38.
*The conditions in each category are to be viewed as selected examples, not comprehensive definitions of the category.

ciation between occupational exposure and disease established on clinical grounds alone.

The typically long latency period between occupational exposure and appearance of illness is another barrier to diagnosis. Added to this is the fact that many workers have had multiple toxic exposures and, at least until recently, have not been informed of the nature or hazard of the materials with which they worked.

Finally, the underdiagnosis of occupational disease reflects the fact that most physicians are not adequately trained to suspect work as a cause of disease. Very little time is devoted in most medical schools to teaching physicians to take a proper occupational history, to recognize the symptoms of common industrial toxins, or to recall known associations between occupational exposure and disease.[42] The average medical student receives only six hours of training in occupational medicine during the four years of medical school.[43] Largely as a result of this lack of training, most physicians do not routinely obtain histories of occupational exposure from their patients. Surveys indicate that adequate occupational histories are recorded on fewer than 10 percent of medical charts.[42]

A major challenge to the medical practitioner is the proper diagnosis of occupationally induced disease.[42,44] Not only will correct diagnosis assist in the treatment and (if necessary) the compensation of the individual patient; it also should serve as a trigger for public health interventions to prevent disease in other similarly exposed workers.[45]

Prevention of Occupational Disease

Primary prevention of occupational disease requires elimination or reduction of hazardous exposures. Secondary prevention depends on the ability to effectively identify potential work-related illness through screening workers at high risk. Tertiary prevention—the prevention of complications and disability resulting from existing illness—depends on the development and application of appropriate diagnostic techniques for identification of persons with established occupational illness. Prevention on all three levels requires solid information on the potential effects of specific occupational exposures, as well as data on the industries and occupations in which hazardous substances are used.

The most effective prevention strategy is the primary prevention of exposure. Reduction in exposure may be accomplished by the following techniques, listed in descending order of preference.

Substituting a less hazardous material is the most efficacious method of controlling a workplace hazard. Substitution of a less hazardous process or piece of equipment also represents a meaningful control strategy. For example, substitution of a continuous process for an intermittent process almost always results in a decrease in exposure. A good example is the operation of solid waste incinerators that release more pollutants during start-up and stopping than in continuous operation.

Engineering controls—such as ventilation, process isolation, or enclosure—may be used to reduce worker exposure to toxic substances. Ventilation is one of the most effective and widely used control measures.

Alteration of work practices can help to reduce exposure to hazards. An example is wet-sweeping asbestos dust rather than dry-sweeping.

Administrative controls are methods of controlling total worker exposure by job rotation, work assignment, or time periods spent away from the hazard. With administrative controls, the level of exposure to the hazard is not diminished; instead, the duration of individual exposure is reduced and exposure is spread more widely among the workforce. The most common use of administrative controls is to reduce overall noise exposure through rotation.

Programs for encouraging personal hygiene constitute another approach to reducing exposure. In some instances, the employer may encourage or require showers and change to clean clothes at the end of the workday. Naturally, the employer should provide showers, changing facilities, lockers, and work clothes if such measures are indicated.

Use of personal protective equipment, such as respirators, gloves, protective clothing, and ear plugs or muffs, can play an important role if carefully designed programs are in place and if the equipment is checked regularly. It is important, however, to recognize that programs of personal protection never constitute as efficient or acceptable a means of protection as engineering or process controls.

Research in the years ahead will be directed toward the development of biological markers of both exposure and disease.[46] Biological markers of exposure, such as the blood-lead level, provide a means for individualized assessment of exposure. Biological markers of disease or of subclinical dysfunction are intended to detect early preclinical pathophysiological changes caused by toxic agents. The use of such markers in worker screening and in population studies may be expected to permit earlier identification of adverse effects when they are at an early and still potentially reversible stage.

Furthermore, the role of the primary-care physician in the recognition and management of occupational diseases must be increased dramatically. The American Board of Medical Specialties identifies only slightly more than 1,000 physicians in the United States today who are board certified in occupational medicine. Of the more than 104 million individuals in the workforce, approximately 70 percent work in facilities with no medical services. These numbers demonstrate that the only alternative for working men and women is to rely on their primary-care physicians for the recognition and management of potential work-related disorders that may be attributable to hazardous exposures encountered at work.[47]

Needed Federal Programs

Child-protective environmental practices and standards are needed in the home, schools, and industry. Changes in governmental rules and regulations are a necessary first step.

Children's Environmental Health

A major need in children's environmental health includes the enactment of a legislative agenda that would support the research and prevention programs needed to protect children from environmental health threats now and in the future. The principal item recommended for passage is a comprehensive Children's Environmental Protection Act, which has the potential to create a broad environmental safety net for children that would cut across and tie together the current patchwork of environmental laws dealing with air, food, water, pesticides, and hazardous waste sites. A comprehensive Children's Environmental Protection Act should include three critical elements. First, it should include a requirement for the provision of information to parents about potential environmental hazards to children, also known as "Right-to-Know." Second, every chemical ought to be thoroughly tested to ensure that it is not hazardous to children or to human development before being introduced to the market. Third, the act should include a legally enforceable requirement mandating that all federal, environmental, and food protection standards be set at levels that would protect the health of infants and children.

In addition to the passage of a Children's Environmental Protection Act, support should also continue for the following six programs:

The National Longitudinal Study

This is a major prospective epidemiologic study proceeding under the direction of the National Institute of Child Health and Human Development, with additional support from the National Institute of Environmental Health Sciences, the Centers for Disease Control and Prevention (CDC), and the U.S. Environmental Protection Agency. The study will examine the effects of early environmental exposures on children's health and development.

Centers for Children's Environmental Health and Prevention Research

These eight centers, supported jointly by the National Institute of Environmental Health Sciences and the U.S. Environmental Protection

Agency, promote translation of basic research findings into applied intervention and prevention methods.

Disease and Exposure Tracking
It will be extremely important to continue to provide support to the CDC so that each year it can continue to monitor the levels of chemicals in the blood of Americans and to make this information available to the public. Also, it is important that the CDC be appropriated sufficient funds to track critically important diseases in children, including asthma, birth defects, and learning disabilities, which is essential for the recognition of clusters of disease.

Pediatric Environmental Health Specialty Units
The Agency for Toxic Substances and Disease Registry (ATSDR) currently supports ten pediatric environmental health specialty units (PEHSUs) at medical schools and children's hospitals across the United States. These PEHSUs are clinical centers that see children with disease of environmental origin and children who have sustained toxic environmental exposures.

National Fellowship Training Program in Environmental Pediatrics
A new program that will be launched within the next year is a national fellowship training program that will train pediatricians in environmental pediatrics. A pilot fellowship has been established by the Ambulatory Pediatric Association, a professional society of academic pediatricians, with the support of private foundations. To sustain this effort in the years ahead and to produce the next generation of leaders in environmental pediatrics, it is important that federal funds be provided to support this program.

The Office of Children's Health Protection (OCHP) at the U.S. Environmental Protection Agency
It is important that OCHP be provided sufficient funds in the years ahead to fulfill its mandate. The work of OCHP will become particularly important if a Children's Environmental Protection Act is passed into law.

Occupational Health

A major future need in occupational health will be for much more widespread use of premarket testing of chemical substances. The widespread lack of toxicity data on chemicals in commerce and the current inadequacies in premarket evaluation of the toxicity of newly developed chemical compounds will need to be corrected.[48]

The lack of a national occupational health surveillance system in the United States is a serious impediment to disease prevention. Surveillance, also known as disease and exposure tracking, is a technique for the systematic recognition and tabulation of occupational hazards and diseases.[49] Hazard surveillance provides a means of assessing toxic occupational exposures and thus of assessing risk.[50] Occupational disease surveillance provides a means of assessing the amount and types of occupational diseases, their time trends, and their distribution according to geography, industry, and occupation. These two types of surveillance complement each other, each being an integral component of a complete occupational health surveillance system.

References

1. Haggerty R, Rothmann J, Press IB. *Child health and the community.* New York: Wiley, 1975.

2. DiLiberti JH, Jackson CR. Long-term trends in childhood infectious disease mortality rates. *Amer J Public Health* 1999; 89:1883–1889.

3. Centers for Disease Control and Prevention. Asthma—United States, 1982–1992. *MMWR* 1995; 43:952–955.

4. DeVesa SS, Blot WJ, Stone BJ, Miller BA, Tarove RE, Fraumeni JF Jr. Recent cancer trends in the United States. *JNCI* 1995; 87:175–182.

5. Zahm SH, Devesa SS. Childhood cancer: Overview of incidence trends and environmental carcinogens. *Environ Health Perspect* 1995; 103 Suppl 6:177–184.

6. Buxbaum L, Boyle C, Yearginn-Allsopp M, Murphy CC, Roberts HE. *Etiology of mental retardation among children ages 3–10: The Metropolitan Atlanta Developmental Disabilities Surveillance Program.* Atlanta: Centers for Disease Control and Prevention, 2000.

7. Kiely M. The prevalence of mental retardation. *Epidemiol Rev* 1987; 9:194–218.

8. American Academy of Pediatrics. Committee on Environmental Health. Environmental tobacco smoke: A hazard to children. *Pediatrics* 1997; 99:639–642.

9. Bearer C, Emerson RK, O'Riordan MA, Roitman E, Shackleton C. Maternal tobacco smoke exposure and persistent pulmonary hypertension of the newborn. *Environ Health Perspec* 1997; 105:202–206.

10. Golding J. Sudden infant death syndrome and parental smoking—A literature review. *Paediatr Perinat Epidemiol* 1997; 11:67–77.

11. Paulozzi LJ, Erickson JD, Jackson RJ. Hypospadias trends in two US surveillance systems. *Pediatrics* 1997; 100:831–834.

12. Schaffer, M. Children and toxic substances: Confronting a major public health challenge. *Environ Health Perspect* 1994; 102 Suppl 2:155–156.

13. US Environmental Protection Agency. *Environmental threats to children's health.* Washington, DC: US Environmental Protection Agency, 1996.

14. Patandin S, Lanting CI, Mulder PG, Boersma ER, Sauer PJ, Weisglas-Kuperus N. Effects of environmental exposure to polychlorinated biphenyls and dioxins on cognitive abilities in Dutch children at 42 months of age. *J Pediatr* 1999; 134(1):33–41.

15. US Environmental Protection Agency (1997, July). *Chemicals-in-commerce information system.* Chemical Update System Database, 1998.

16. National Academy of Sciences. *Toxicity testing: Needs and priorities.* Washington, DC: National Academy Press, 1984.

17. US Environmental Protection Agency, Office of Pollution Prevention and Toxic Substances. *Chemical hazard data availability study: What do we really know about the safety of high production volume chemicals?* Washington, DC: US Environmental Protection Agency, 1998.

18. National Research Council. *Pesticides in the diets of infants and children.* Washington, DC: National Academy Press, 1993.

19. Landrigan PJ, Suk WA, Amler RW. Chemical wastes, children's health and the Superfund Basic Research Program. *Environ Health Perspect* 1999; 107:423–427.

20. Charnley G, Putzrath RM. Children's health, susceptibility, and regulatory approaches to reducing risks from chemical carcinogens. *Environ Health Perspect* 2001; 109:187–192.

21. Gibson JL. A plea for painted railings and painted walls of rooms as the source of lead poisoning among Queensland children. *Aus Med Gazette* 1904; 23:149–153.

22. Miller RW. Delayed effects occurring within the first decade after exposure of young individuals to the Hiroshima atomic bomb. *Pediatrics* 1956; 18:1–18.

23. Wakeford R. The risk of childhood cancer for intrauterine and preconceptional exposure to ionizing radiation. *Environ Health Perspect* 1995; 103:1018–1025.

24. Miller RW, Blot WJ. Small head size after exposure to the atomic bomb. *Lancet* 1972; 2:784–787.

25. Harada H. Congenital Minamata disease: Intrauterine methylmercury poisoning. *Teratol* 1978; 18:285–288.

26. Landrigan PJ. The toxicity of lead at low dose. *Brit J Industr Med* 1989; 46:593–596.

27. Needleman HL, Gunnoe C, Leviton A. Deficits in psychological and classroom performance of children with elevated dentine lead levels. *N Engl J Med* 1979 300:689–695.

28. Bellinger D, Leviton A, Waternaux C, Needleman H, Rabinowitz M. Longitudinal analysis of prenatal and postnatal lead exposure and early cognitive development. *N Engl J Med* 1987; 316:1037–1043.

29. Jacobson JL, Jacobson SW, Humphrey HEB. Effects of in utero exposure to polychlorinated biphenyls and related contaminants on cognitive functioning in young children. *J Pediatr* 1990; 116:38–45.

30. Grandjean P, Weihe P, White RF, Debes F, Araki S, Yokoyama K, et al. Cognitive deficit in 7-year-old children with prenatal exposure to methylmercury. *Neurotoxicol Teratol* 1997; 19:417–428.

31. Landrigan PJ. *The burden of occupational disease in the United States.* Testimony before the Committee on Labor and Human Resources, United States Senate, Oversight Hearings on the Occupational Safety and Health Act (April 18, 1988).

32. Baker EL, Folland DS, Taylor TA, Frank M, Peterson W, Lovejoy G, et al. Lead poisoning in children of lead workers: Home contamination with industrial dust. *N Engl J Med* 1977; 296:260–261.

33. Chisolm JJ Jr. Fouling one's own nest. *Pediatrics* 1978; 62:614–617.

34. Levy BS, Wegman DH, editors. *Occupational health-recognizing and preventing work-related diseases.* 2nd ed. New York: Little, Brown, 1988.

35. Rom WN, ed. *Environmental and occupational medicine.* 2nd ed. New York: Little, Brown, 1992.

36. Zenz, C. *Occupational medicine: Principles and practical application.* 2nd ed. Chicago: Yearbook Medical Publications, 1988.

37. National Institute for Occupational Safety and Health. *Proposed national strategies for the prevention of leading work-related diseases and injuries—Part 1.* Washington, DC: Association of Schools of Public Health, 1986.

38. National Institute for Occupational Safety and Health. *Proposed national strategies for the prevention of leading work-related diseases and injuries—Part 2.* Washington, DC: Association of Schools of Public Health, 1988.

39. Selikoff IJ, Churg J, Hammond EC. Asbestos exposure and neoplasia. *JAMA* 1964; 188:22–26.

40. Landrigan PJ. Toxic exposures and psychiatric disease—Lessons from the epidemiology of cancer. *Acta Psychiatrica Scandinavica* 1983; 67 Suppl 303:6–15.

41. Creech JL Jr and Johnson, MN. Angiosarcoma of the liver in the manufacture of polyvinyl chloride. *J Occup Med* 1974; 16:150–151.

42. Institute of Medicine. *Role of the primary care physician in occupational and environmental medicine.* Washington, DC: National Academy Press, 1988.

43. Burstein JM, Levy BS. The teaching of occupational health in US medical schools: Little improvement in 9 years. *Am J Public Health* 1994; 84(5):846–849.

44. *The role of the internist in occupational medicine.* Philadelphia: American College of Physicians, 1984.

45. Rutstein DD, Mullan RJ, Frazier TM, Halperin WE, Melius JM, Sestito JP. Sentinel health events (occupational): A basis for physician recognition and public health surveillance. *Am J Public Health* 1983; 73:1054–1062.

46. Committee on Biological Markers of the National Research Council. Biological markers in environmental health research. *Environ Health Perspect* 1987; 74:3–9.

47. Rosenstock L. Occupational medicine: Too long neglected. *Ann Intern Med* 1981; 95: 664–676.

48. National Research Council. *Toxicity testing: Strategies to determine needs and priorities*. Washington, DC: National Academy Press, 1984.

49. Langmuir, AD. The surveillance of communicable diseases of national importance. *N Engl J Med* 1963; 268:182–192.

50. Wegman DH, Froines JR. Surveillance needs for occupational health. *Am J Public Health* 1985; 75:1259–1261.

War and the Environment

Jennifer Leaning

16

War has marked human experience since the beginning of recorded time, and the demands of war have in many ways shaped and advanced the practice of medicine.[1,2] Rhodes has estimated the immense scope of war-related mortality in the twentieth century and demonstrated the increasing fraction of civilian deaths.[3] Levy and Sidel have reviewed the broad public health consequences of preparing for, coordinating, and cleaning up after contemporary wars.[4] War rivals infectious disease as a global cause of morbidity and mortality. In the 1980s, health professionals' concern about the effects of war on the environment[5] was focused on the sweeping ecological consequences of nuclear weapons.[6] In the 1990s, the Gulf War and the Kosovo experiences demonstrated the environmentally destructive capacities of conventional weapons.[7]

Enormous gaps remain in our knowledge about the relationship of war and health. Understanding is constrained by the lack of recorded information and the comparative absence of continuing systematic field research undertaken from within any one discipline. Work is underway to explore how environmental stress and resource constraints may contribute to conflict,[8,9] but the topic lies outside the scope of this chapter.

The environmental impacts of war can be understood by examining the magnitude and duration of effects, involved ecosystems in specified geographic locations, the use of individual weapons systems, the results of particular production processes, and the cumulative combined effects of specified military campaigns. From this perspective, four activities can be seen as having prolonged and pervasive environmental impact with significant consequences for human populations: production and testing of nuclear weapons, aerial and naval bombardment of terrain, dispersal and persistence of land mines and buried ordnance, and use or storage of military despoliants, toxins, and waste.

Production and Testing of Nuclear Weapons

Nuclear-weapons technology was developed during World War II and expanded as an industrial enterprise of vast scope and complexity in the Cold War between the United States and the Soviet Union. Nuclear-weapons technology continues to dominate concerns regarding potential hazards to the environment.[10] Radioactivity, released into the environment in many phases of production and testing processes, poses a serious threat to the health of biological species, including humans. (See chapter 12.) Assessment of this threat begins with estimating the amount of radiation released, itself a difficult task, and then evaluating health risks on the basis of what can be found in epidemiologic studies of exposed populations and ecosystems over time. These studies are based on relatively small samples and look at areas affected by aboveground tests,[11–15] areas near nuclear-weapons production and storage facilities,[16] and areas used for radioactivity tests.[17] The studies raise concerns in terms of human health effects, costs of environmental cleanup, and continued environmental contamination.[18–23]

Massive amounts of radioactivity were released in the last half of the twentieth century from the nuclear-weapons testing programs of all the main nuclear powers. The testing phase of nuclear weapons included 423 atmospheric tests (conducted from 1945 to 1957) and about 1,400 underground tests (from 1957 to 1989). The total burden of radionuclides released from these tests has been estimated at 16–18 million curies (1 Ci = 3.7×10^{11} Bq) of strontium-90, 25–29 million curies of cesium-137, 400,000 curies of plutonium-329, and (for the atmospheric tests only) 10 million curies of carbon-14.[11,21]

More is known from the United States than from other countries about radiation releases from the military production of nuclear weapons. Production sites that have been investigated and found to have caused significant environmental contamination include the Hanford Nuclear

Reservation in Washington State (producing weapons-grade plutonium), the Oak Ridge Reservation in Tennessee (producing components for nuclear weapons), the Rocky Flats Plant in Colorado (producing plutonium triggers for warheads), and the Savannah River Plant in Georgia (producing tritium and plutonium). Accidental releases and continued emissions as part of daily operations have been reported at these and many other production facilities.[21] Disputes regarding the human health effects of these exposures have not been entirely resolved, despite extensive study.[24,25] The U.S. government has recently acknowledged that occupational exposures to nuclear and other toxic materials at these plants justifies the awarding of compensation to over 3,000 current and retired workers whose health has been adversely affected.[26]

Aerial and Naval Bombardment

Bombardment of the urban infrastructure, which constitutes the environment for a significant fraction of the world's human population, has always caused forced dislocation of survivors. During World War II, when air power for the first time was deployed as the pivotal military technology, the practice of bombing civilian settlements became increasingly prevalent, and hundreds of thousands of people died as a result.[27] In the aerial bombardments of Tokyo in March 1945, about 100,000 to 200,000 people were killed. In the fire bombings of 70 German cities, including Hamburg in 1943 and Dresden in 1945, it is estimated that 500,000 to 800,000 people died.[28] About 200,000 people died from the acute effects of the atomic bombs in Hiroshima and Nagasaki in 1945.[29]

The bombardment of cities and the destruction of forests, farms, transport systems, and irrigation networks during World War II produced devastating environmental consequences,[30] and by the end of the war there were almost 50 million refugees and displaced people.[31,32] In the last year of the war the land of coastal and northern France was torn up, Holland south of the Zuyder Sea was flooded with the destruction of dikes, and many ports were clogged with unexploded ordnance and sunken ships. Great damage had been done to most cities in Europe, with the hardest hit including Warsaw, Berlin, Hamburg, Dresden, Düsseldorf, Boulogne, Le Havre, Rouen, Brest, Pisa, Verona, Lyons, Budapest, Leningrad, Kiev, and Kraców.

All visitors to central Europe reported a feeling of unreality. What appeared to be lunar landscapes were dotted with enormous heaps of rubble and bomb craters, deserted and stinking ruins that had once been business centers and residential areas. Finding housing for the survivors was the most urgent problem, but in Germany about a quarter of all houses were uninhabitable, and almost as many in Poland, Greece, Yugoslavia, and the European part of the Soviet Union. In the American zone of Germany, 81 percent of all houses had been destroyed or damaged. In the German-occupied parts of the Soviet Union, the homes of 6 million families had been destroyed, leaving about 25 million people without shelter.[33]

Estimates of war damage in Japan noted that 66 cities had suffered major damage, with about 40 percent of their area destroyed; throughout Japan, about 9 million people were left homeless. Comprehensive data are not available, but limited evidence from the first two postwar years suggests that, because of vast food shortages and the failure of the 1945 rice harvest, hunger and malnutrition afflicted the majority of the population, and thousands died from causes related to starvation.[34]

This sequence of aerial bombardment, destruction of home and urban and rural infrastructure, and progressive waves of dislocated or homeless people can be seen in all wars subsequent to World War II. In the 15 years of the war in Southeast Asia, the U.S. bombardment of Vietnam, Laos, and Cambodia forced about 17 million people to become refugees.[35] In the Gulf War, the allied forces crippled the urban support systems of major cities in Iraq.[36–38] In the conflicts of the post–Cold War era, marked by sieges of cities, attacks on safe havens, and pulverization of towns to effect ethnic cleansing, millions of people have been forced to flee within or across national borders. In 1999, about 35 million people were counted as refugees or internally displaced people as a result of war or internal crisis.[39,40]

Land Mines

As a result of the last 50 years of wars in Europe, Africa, Asia, and Latin America, an estimated 70–100 million antipersonnel land mines are still active and in place worldwide, and another 100 million exist in stockpiles.[41] Almost 400 million have been strewn across continents since

World War II, and with the proliferation of civil wars waged by irregular forces, the use and spread of land mines as a preferred method of securing and denying land has accelerated.[42] Land mines are placed now without regard to requirements under international law to mark, map, monitor, and remove them.[43,44] Hence, the majority of the victims of land-mine explosions are civilians engaged in daily farming or foraging activities.[45–48] Reliable regional estimates of incident rates of injury and death are difficult to come by; one frequently cited statistic is that land mines injure or kill about 500 people every week.[49]

The countryside of Kosovo was rimmed and internally laced with land mines laid by all sides; after a year of international efforts to remove them, an estimated 1,415 known or suspected minefields remain. Since the June 1999 ceasefire and the return of the civilian population, the monthly toll killed from land-mine or cluster-bomb explosions has dropped from 44 deaths and 109 serious injuries in June 1999 to no deaths and 15 serious injuries in April 2000.[50]

Land mines accelerate environmental damage through one of four mechanisms: fear of mines denies access to abundant natural resources and arable land; populations are forced to move preferentially into marginal and fragile environments to avoid minefields; this migration speeds depletion of biological diversity; and land-mine explosions disrupt essential soil and water processes.

Review of experiences in the twentieth century indicates that the persistence of active mines and unexploded ordnance haunts old battle areas and that, despite intensive efforts at clearance and deactivation, millions of hectares remain under interdiction in Europe, North Africa, and Asia.[51] In Libya, one-third of the land mass is considered contaminated by land mines and unexploded munitions from World War II.[52] When these mines do explode, in addition to causing serious injury and death to humans, domestic animals, and wildlife, they shatter soil systems, destroy plant life, and disrupt water flows, accelerating ecosystem disruption.[53] Interactions between natural disasters and buried land mines slow attempts to de-mine areas and protect populations. For instance, the floods in Mozambique in 1999 and 2000 are feared to have displaced the hundreds of thousands of land mines left from the civil war, and concern about their whereabouts has delayed recovery operations. Painstaking efforts to mark known minefields have been set back considerably by the floodwaters, and a new mapping team has been sent out by the international community.[54]

Despoliation, Defoliation, and Toxic Pollution

Attempts to damage the environment as a tactic of war against the formal enemy and as a means of instilling terror in the general populace have been described throughout history.[55,56] During World War II, instances of dike disruption,[57,58] dam destruction,[59] and scorched-earth retreats[60] have been well documented. Interactions between natural disasters and buried land mines slow attempts to de-mine areas and protect populations.

It is generally accepted that the extensive use of environmental destruction as a strategic practice in war dates from the use of defoliants during the war in Southeast Asia. From 1965 to 1971, the United States sprayed 3,640 km^2 of South Vietnam's cropland with herbicides, using a total estimated amount of 55 million kg. The stated rationale was to deny the enemy sources of food and means of cover.[35] This widespread use of chemicals to destroy farmland, forest, and water sources is unprecedented, and the environmental consequences are still relatively unexplored. Early observations raised concerns about environmental effects,[61] but more comprehensive studies had to await cessation of hostilities. International teams have been granted access for field assessments only in the last few years.

Of the many wars waged since Vietnam, the Gulf War during January and February 1991 demonstrates the ways in which the technologies of war and industry can be used to wreak widespread environmental havoc. Iraq's release of about 10 million barrels of Kuwaiti oil into Gulf waters[62] caused great stress to an ecosystem already suffering from decades of abuse (oil spills, the Iraq-Iran war, freighter traffic, and industrial waste). Scientific assessments of this ecological loss and the catastrophe resulting from the Iraqi firing of 732 Kuwaiti oil wells are underway,[63,64] although constrained by incomplete data and controversy.[65–69]

More recent wars, or what the humanitarian relief community terms "complex humanitarian emergencies,"[70,71] have been assessed for their potential, through the creation of large refugee camps, to inflict harm on the local environment in which the camps are situated. In the cases of the refugee camps in the African Great Lakes region from 1994 to 1997

(Mozambique, Sudan, and the Afghanistan-Pakistan border areas), a number of studies are now looking at issues of deforestation, encroachment on vulnerable ecosystems and national parks, water pollution and sanitation degradation, air pollution, and loss of endangered species.[72,73]

Prescriptions

Future work on the environmental effects of war must address four main issues: information, threat assessment, vulnerability assessment, and the role of international law.

Information

Insufficient information exists about the effects of war on natural ecosystems, both in the immediate aftermath of war and over the long term. Methods for historical and contemporaneous reporting are incompletely developed and lack robust institutional support. Without improvement in these areas, assessments of the environmental damage of war will continue to be fragmentary.

Threat Assessment

Escalation in numbers of weapons, advances in technology, and widespread proliferation, including threats of terrorist use,[74,75] now place the local, regional, and global environment in greater jeopardy than ever before. Nuclear weapons, the most extreme technology, have been shown in careful theoretical studies to be capable, even in limited regional use, of destroying vast sections of the world's environment.[76] Despite the fact that our capacity to contain and mitigate environmental effects of current weapons systems used in war is grossly underdeveloped, the world community continues to permit, and even support, a multiplicity of regional and international arms races.[77]

Vulnerability Assessment

Historical data on the destruction of coral reefs during the war in the Pacific[78] and enduring changes in desert terrain from the North African campaigns of World War II[79,80] provide faint and isolated hints that fragile environments take a long time to recover from war. Burdened by rapid population growth in many parts of the world, unrestrained settlement, and economic exploitation, regional ecosystems are increasingly threatened.[81] As we encroach on the margins of our environment in the twenty-first century, postwar ecological resilience cannot be assumed to be present in all places, particularly within a human timeframe.

International Law

The legal and ethical framework within which the medical and public health profession works during wartime is defined by the Geneva Conventions and related documents. The current discussion is whether existing law to limit the environmental effects of war is sufficient, if fully enforced, or whether new law is needed. Proposals to set up environmental surveillance systems, as enforcement mechanisms to support current law, were developed during the Gulf War.[82–84]

The increasing participation of health care professionals in these four areas of work may lend impetus to the development of research and policy leading to more positive outcomes.

References

1. Gabriel RA, Metz KS. *A history of military medicine.* New York: Greenwood Press, 1992.

2. Taliaferro WH, editor. *Medicine and the war.* Chicago: University of Chicago Press, 1972.

3. Rhodes R. Man-made death: A neglected mortality. *JAMA* 1988; 260:686–687.

4. Levy BS, Sidel VW. *War and public health.* New York: Oxford University Press, 1997.

5. Leaning J. War and the environment: Human health consequences of the environmental damage of war. In: Chivian E, McCally M, Hu H, Haines A, editors. *Critical condition: Human health and the environment.* Cambridge, MA: MIT Press, 1993. pp. 123–137.

6. Barnaby F. The spread of the capability to do violence: An introduction to environmental warfare. *Ambio* 1975; 4:178–244.

7. Finger M. The military, the nation state, and the environment. *Ecologist* 1991; 21:220–225.

8. Homer-Dixon TF. On the threshold: Environmental change as causes of acute conflict. *Int Security* 1991; 16:76–116.

9. Homer-Dixon TF. Environmental scarcities and violent conflict: Evidence from cases. *Int Security* 1994; 19:5–40.

10. Renner M. Assessing the military's war on the environment. In: Brown L, editor. *State of the world 1991.* New York: WW Norton, 1991. pp. 132–152.

11. International Physicians for the Prevention of Nuclear War and the Institute for Energy and Environmental Research. *Radioactive heaven and earth: The health and environmental effects of nuclear weapons testing in, on, and above the earth*. New York: Apex Press, 1991.

12. Danielsson B. Poisoned Pacific: The legacy of French nuclear testing. *Bull Atomic Sci* 1990; 46(2):22–34.

13. An end to all nuclear explosions: The long-overdue test ban. *Defense Monitor* 1991; 20(3):1–8.

14. Lyon JL, Klauber MR, Gardner JW, Udall KS. Childhood leukemias associated with fallout from nuclear testing. *N Engl J Med* 1979; 300:397–402.

15. Hamilton TE, van Belle G, LoGerfo JP. Thyroid neoplasia in Marshall Islanders exposed to nuclear fallout. *JAMA* 1987; 258:629–636.

16. Makhijani A, Hu H, Yih K, editors. *Nuclear wastelands: A global guide to nuclear weapons production and its health and environmental effects*. Cambridge, MA: MIT Press, 1995.

17. Bensen DW, Sparrow AH. Survival of food crops and livestock in the event of nuclear war. In: *Proceedings of symposium at Brookhaven National Laboratory*. Washington, DC: US Atomic Energy Commission, 1971.

18. Cochran TB, Norris RS. A first look at the Soviet bomb complex. *Bull Atomic Sci* 1991; 47(4):25–31. Available: www.bullatomsci.org/issues/1991/may91/may91 cochran.html

19. Kitfield J. The environmental cleanup quagmire. *Military Forum* 1989; Apr:36–40.

20. Coyle D. *Deadly defense: Military radioactive landfills*. New York: Radioactive Waste Campaign, 1988.

21. Office of Technology Assessment. *Complex cleanup: The environmental legacy of nuclear weapons production*. Washington, DC: Office of Technology Assessment, 1991. Report No: OTA–0–484. Available: www.wws.princeton.edu/~ota/ns20/alpha_f.html

22. National Cancer Institute. *Study estimating thyroid doses of I–131 received by Americans from Nevada atmospheric nuclear bomb tests*. Washington, DC: National Cancer Institute, 1997.

23. Sidel VW, Shahi G. The impact of military activities on development, environment, and health. In: Shahi G, Levy B, Binger A, Kjellstrom T, Lawrence R, editors. *International perspectives on environment, development, and health: Toward a sustainable world*. New York: Springer, 1997. pp. 283–312.

24. Geiger HJ, Rush D. *Dead reckoning: A critical review of the Department of Energy's epidemiologic research*. Washington, DC: Physicians for Social Responsibility, 1992.

25. Committee on the Biological Effects of Ionizing Radiation, National Research Council. *Health effects of exposure to low levels of ionizing radiation*. BEIR V. Washington, DC: National Academy Press, 1990.

26. *Secretary Richardson announces proposal to compensate thousands of sick workers* [press release]. Washington, DC: US Department of Energy, 2000 Apr 12. Available: www.doe.gov/news/releases00/aprpr/pr00103.htm

27. Westing AH. Misspent energy: Munition expenditures past and future. *Bull Peace Proposals* 1986; 16(1):9–10.

28. Postel T. A review of the physics of large urban fires. In: Solomon F, Marston RQ, editors. *The medical implications of nuclear war.* Washington, DC: National Academy Press, 1986. pp. 73–95.

29. Committee for the Compilation of Materials Caused by the Atomic Bombs in Hiroshima and Nagasaki. *Hiroshima and Nagasaki: The physical, medical, and social effects of the atomic bombings.* New York: Basic Books, 1981.

30. *United States strategic bombing survey.* New York: Garland, 1976.

31. Laquer W. *Europe since Hitler.* New York: Penguin Books, 1984.

32. Proudfoot MJ. *European refugees 1939–1952: A study in forced population movement.* Evanston, IL: Northwestern University Press, 1956.

33. Laquer W. *Europe since Hitler.* New York: Penguin Books, 1984.

34. Dower JW. *Embracing defeat: Japan in the wake of World War II.* New York: WW Norton, 1999.

35. Westing AH. *Warfare in a fragile world: Military impacts on the human environment.* London: Taylor and Francis, 1980.

36. Arkin WM, Durrant D, Cherni M. *On impact: Modern warfare and the environment; a case study of the Gulf War.* London: Greenpeace, 1991.

37. Save the Children Fund. *Iraq situation report for Save the Children.* London: Save the Children Fund, 1991.

38. Renner MG. *Military victory, ecological defeat.* World Watch 1991; July/Aug:27–33.

39. US Committee for Refugees Web site. Available: www.refugees.org

40. United States Mission to the UN. *Global humanitarian emergencies 1998–99.* New York: United Nations, 1999.

41. Department of Humanitarian Affairs, Mine Clearance and Policy Unit. *Land mine facts.* New York: United Nations, 1996.

42. Westing AH. *Warfare in a fragile world: Military impact on the human environment.* London: Taylor and Francis, 1980.

43. Arms Project of Human Rights Watch and Physicians for Human Rights. *Landmines: A deadly legacy.* New York: Human Rights Watch, 1993.

44. Prokosch E. *The technology of killing: A military and political history of antipersonnel weapons.* London: Zed Books, 1995.

45. Coupland R, Korver A. Injuries from anti-personnel mines: The experience of the International Committee of the Red Cross. *Br Med J* 1991; 303:1509–1512.

46. Stover E, Keller AS, Cobey J, Sopheap S. The medical and social consequences of land mines in Cambodia. *JAMA* 1994; 272:331–336.

47. Strada G. The horror of land mines. *Sci Am* May1996; 274:40–45.

48. Ascherio A, Biellik R, Epstein A, Snetro G, Gloyd S, Ayotte B, et al. Deaths and injuries caused by land mines in Mozambique. *Lancet* 1995; 346:721–724.

49. Giannou C. Antipersonnel landmines: Facts, fictions, and priorities. *BMJ* 1997; 315:1453–1454.

50. International Crisis Group. *Kosovo report card.* Pristina/Brussels: International Crisis Group, 2000 Aug 28. ICG Balkans Report No: 100.

51. Westing AH, editor. *Explosive remnants of war: Mitigating the environmental effects.* London: Taylor and Francis, 1985.

52. Sgaier K. Explosive remnants of World War II in Libya: Impact on agricultural development. In: Westing AH, editor. *Explosive remnants of war: Mitigating the environmental effects.* London: Taylor and Francis, 1985. pp. 33–37.

53. Eliasson J. An international approach towards humanitarian assistance and economic development of countries affected by land mines. In: Cahill K, editor. *Clearing the fields: Solutions to the global land mines crisis.* New York: Basic Books, 1995. pp. 308–318.

54. International Campaign to Ban Landmines. *Landmine monitor report 1999: Toward a mine-free world.* New York: Human Rights Watch, 1999.

55. Jones A. *The art of war in the western world.* New York: Oxford University Press, 1987.

56. Royster C. *The destructive war: William Tecumseh Sherman, Stonewall Jackson, and the Americans.* New York: Knopf, 1991.

57. Westing AH, editor. *Environmental hazards of war: Releasing dangerous forces in an industrialized world.* Thousand Oaks, CA: Sage, 1990.

58. Westing AH. *Weapons of mass destruction and the environment.* London: Taylor and Francis, 1977.

59. The attack on the irrigation dams in North Korea. *Air Univ Q Rev* 1953/54; 6 (winter):40–61.

60. Lund DH. The revival of northern Norway. *Geographical J* 1947; 109:185–197.

61. Orians GH, Pfeiffer EW. Ecological effects of the war in Vietnam. *Science* 1970; 168:544–554.

62. United Nations Environment Programme. Introductory report of the Executive Director. *Environmental consequences of the armed conflict between Iraq and Kuwait.* New York: UNEP, 1991 May 10. Report No: UNEP/GC.16/4/Add.1.

63. Carothers A. *After Desert Storm: The deluge.* Greenpeace; Oct/Nov/Dec 1991. pp. 14–17.

64. Sheppard C, Price A. Will marine life survive the Gulf War? *New Scientist* 1991; Mar 9:36–40.

65. Horgan J. Up in flames. *Sci Am* 1991; 264(5):17–24.

66. Horgan J. Burning questions. *Sci Am* 1991; 265(1):17–24.

67. Wald ML. Experts worried by Kuwait fires. *New York Times* 1991 Aug 14. p.A7

68. Stone R. Kuwait quits smoking. *Science* 1992:255:1357.

69. Canby TY. After the storm. *National Geographic* 1991; 180(2):2–34.

70. *Agenda for development.* New York: United Nations, 1997.

71. Kaldor M, Vashee B, editors. *The new wars.* Vol I of *Restructuring the global military sector.* London: UNU World Institute for Development Economics Research, 1997.

72. United Nations High Commissioner for Refugees, International Organization for Migration and Refugee Policy Group. Environmentally-induced population displacements and environmental impacts resulting from mass migration. In: *International symposium report, April 21–24, 1996.* Geneva: International Organization for Migration, 1996.

73. Kibreab G. Environmental causes and impact of refugee movements: A critique of the current debate. *Disasters* 1997; 21:20–38.

74. US Public Health Service, Office of Emergency Preparedness. *Proceedings of the seminar on responding to the consequences of chemical and biological terrorism, 1995 Jul 11–14; Washington, DC.* Washington, DC: Office of Emergency Preparedness, 1996. Report No: 1996-416-003.

75. Tucker JB. National health and medical services response to incidents of chemical and biological terrorism. *JAMA* 1997; 278:362–368.

76. Arkin W, von Hippel F, Levi BG. The consequences of a "limited" nuclear war in East and West Germany. In: Peterson J, editor. *The aftermath: The human and ecological consequences of nuclear war.* New York: Pantheon Books, 1983. pp. 165–187.

77. Renner M. Budgeting for disarmament: The costs of war and peace [*World Watch* paper 122]. Washington, DC: World Watch Institute, 1994.

78. Ruff T. Ciguatera in the Pacific: A link with military activities. *Lancet* 1989; 1:201–205.

79. Oliver FW. Dust-storms in Egypt and their relation to the war period, as noted in Maryut, 1939–1945. *Geographical J* 1945; 106:26–49.

80. Oliver FW. Duststorms in Egypt as noted in Maryut: A supplement. *Geographical J* 1947; 108:221–226.

81. World Commission on Environment and Development. *Our common future.* Oxford: Oxford University Press, 1987.

82. Barnaby F. The environmental impact of the Gulf War. *Ecologist* 1991; 21:166–172.

83. Toukan A. Humanity at war: The environmental price. *Physician Soc Respons Q* 1991; 1:214–220.

84. Nimetz M, Caine GM. Crimes against nature. *Amicus J* 1991; 13:8–10.

Sustainable Health Care and Emerging Ethical Responsibilities

Andrew Jameton and Jessica Pierce

17

The declining condition of the natural environment is beginning to affect the health of populations in many parts of the world.[1–5] As a result, health care professionals and organizations need to consider the long-term environmental costs of providing health care and to reduce the material and energy consumption of the health care industry. This may seem a surprising conclusion, given that average human health has, for the most part, improved in recent decades despite environmental decline. As indicated in the World Health Organization's fiftieth anniversary report,[6] the average life expectancy at birth worldwide has increased rapidly (from 46 years in 1958 to an unprecedented 66 years in 1998), the rate of death among children under five has decreased, more people than ever before have access to at least minimal health care services as well as safe water and sanitation. New vaccinations and medications await wide distribution.

Yet these achievements are fragile. In the long term, human health requires a healthy global ecosystem.[7–11] About 25 percent of health problems are already environmental in origin.[12] There is no realistic way or current technology available to replace declining natural ecosystem services (e.g., climate stabilization, water purification, waste decomposition, pest control, seed dispersal, soil renewal, pollination, biodiversity, and protection against solar radiation) that are essential to health.[13] Although public health experts increasingly recognize the significant role the environment plays in public health, it is less well recognized that personal health care services also depend significantly on and have consequences for the environment.

Linking Health Care and the Environment

Health care figures both as a solution to environmental decline and as a problem. Increasing health problems generated by environmental decline will require medical treatment. At the same time, health care services also damage the environment. In the United States, such services generate over 3 million tons of solid waste per year.[14] As with other service industries such as hotels and restaurants, hospitals consume energy in heating, cooling, manufacturing, and transportation. They occupy large, complex buildings surrounded by concrete and asphalt surfaces; they use high-volume food services, laundry, high-speed transportation, and paper, packaging, and disposable supplies. Health services also pose unique problems, including the use of pharmaceutical and biological products with complex manufacturing processes, environmentally significant precursors, and potentially toxic bodily by-products of medications, as well as complex and hazardous solid, air, and water emissions, including toxic, infectious, and radioactive wastes.[15]

Environmental costs are most evident at the downstream end of health care: the by-products that leave the system as waste. The problems of medical waste, particularly infectious materials (e.g., human tissues and blood) and biohazardous agents (e.g., heavy metals and radioactive isotopes), are fairly well understood and regulated.[16–18] Still, several groups are exploring additional sources of environmental harm. Health Care Without Harm, a coalition of activists and health care organizations, has advocated the elimination of mercury from health care products.[19] They encourage the separation of polyvinyl chloride (PVC) plastics from the collection of infectious wastes because, during the common practice of incinerating infectious wastes, carcinogenic dioxins are released when PVC products are included. Most recently, and more controversially, the coalition has begun warning patients and health care professionals against the use of vinyl intravenous bags containing phthalate plasticizers, which may leach toxins into patients' bodies.[20,21]

The Sustainable Hospitals Project of the Lowell Center for Sustainable Production, in Lowell, Massachusetts, is also working on pollution prevention in health care facilities. It is focusing attention on alternatives to products that contain potentially harmful materials such as latex, PVC, and mercury, particularly through influencing purchasing practices within hospitals.[22] The Green Health Center project at the University of Nebraska Medical Center is exploring the ethical principles relevant to providing environmentally sound, high-quality health care.[23]

Not as well understood, yet perhaps more important than pollution downstream from health care services, are the environmental effects upstream from health care delivery. Health care services rely on an enormous array of natural resources, including common and rare metals, naturally occurring pharmaceutical precursors, rubber, petroleum, biomass, and water. Intravenous pumps, x-ray films, latex gloves—each of these common hospital items requires complex manufacturing processes with attendant environmental effects, many of which are felt on the other side of the world.[24] The environmental costs of natural-resource consumption in health care have not been carefully studied, so the degree to which health care activities contribute to environmental deterioration is difficult to assess. However, because health care services represent a significant sector of intensive North American economies, health care shares responsibility for the environmental problems created by the acquisition, processing, and transportation of natural resources required to make the supplies and energy used by consumers.

Sustainable Health Care

If the earth's ecosystem is to continue to support human health, each community needs to maintain public health and provide health care in ways that will sustain the earth's ecosystem. By many accounts, the environmental crisis results from a combination of population growth, consumption patterns, and technology choices.[25,26] That world population growth must be slowed, and probably reversed, to avoid overwhelming the earth's natural systems has been recognized by many.[27,28] Less well recognized, but equally important, is that the billion members of the world's consumer class threaten future human welfare with their material, and energy-intensive, lifestyles.[29] One way to represent the scale of consumption is to use the ecological "footprint": an estimate of the amount of space it takes to generate the energy, food, pasture, consumer goods, and so on that it takes to maintain each of us. The Ecological Footprints of Nations Study calculates that "humanity as a whole uses over one-third more resources and ecoservices than what nature can regenerate."[30] The United

States has a footprint of 9.6 hectares per capita, whereas Canada's average per capita footprint is 7.2 hectares, still well over the 1.7 hectares globally available per capita.[30]

Large-scale health care systems, such as those in Canada and the United States, depend on wealthy economies to sustain them. But wealthy economies are unsustainable and must scale down their overall consumption of materials and energy.[31–35] If wealthy industrialized societies as a whole are unsustainable, then so are the health care systems housed by these societies.[36] And if the material scale of these economies is to be reduced, so must the scale of health care. The degree of reduction needed is extremely uncertain, but "Factor X" debates in Europe have set goals for overall reduction of national throughput by factors of one-half, one-fourth, one-tenth, and even smaller fractions of current material and energy consumption.[37] These levels of reduction are likely to prove very challenging. Indeed, it is extremely unlikely that any imaginable new technology could achieve these reductions in resource consumption without a substantial reduction in the supply of consumer products as well.[29,38,39] Although no empirical data show that the scale of consumption of natural resources required by industrialized health care systems is ecologically unsustainable, current levels of consumption may challenge our ability to provide health care for future generations. We should thus examine how we can reduce the scale of health care to more modest, sustainable levels.

The environmental impact of health care and the puzzle of sustainability raise ethical questions regarding health care's environmental stewardship. Concern for the health of the earth's ecosystems suggests that health care institutions and practitioners should reassess their practices in order to soften or eliminate harmful effects. At the same time, they have to balance their environmental responsibilities with their obligations to serve the immediate needs of patients. Addressing the issue of balance requires combining considerations from both medical ethics and environmental ethics.

The field of medical ethics has focused largely on principles of human autonomy and issues surrounding benefit to individual patients.[40–42] It has made significant contributions to our understanding of bedside care, decision making with patients, and the use of new health care technologies. Although some bioethicists have discussed issues with environmental implications—such as new technologies, genetic engineering, overpopulation, and treatment of animals—few ethicists have linked bedside concerns to the larger context of global environmental well-being.[43–46] Meanwhile, the field of environmental ethics has grown extensively, but with little attention to medicine and health care. It is time for environmental ethics and medical ethics to reopen a dialogue and seek an ethically appropriate balance between immediate individual health needs and sustainability.

The ethical argument for considering sustainability in health care arises from basic ethical commitments common to environmental and medical ethics. First, today's generations have responsibilities for the welfare of future generations. Along with society at large, health care should accept a responsibility to meet current needs in ways modest and clean enough to be sustainable for centuries.[47,48] Second, humans have a responsibility toward the natural world for the sake of both nature and ourselves.[49] Indeed, action to reduce the impact of humans on nature is urgent; the World Wide Fund for Nature estimates an overall decline of 30 percent in the state of nature since 1970.[50] Third, because about 80 percent of the world's wealth benefits only 20 percent of its people, the vast majority have very little. Poverty is one of the main factors contributing to poor health, and it reduces the ability of populations to cope with environmental decline.[51,52] Justice and sustainability require that health care services be more equitably allocated on a global scale.

But why should the world's wealthy consumer classes, who spend roughly 90 percent of all of the dollars spent on health care in the world,[53] be sensitive to ethical principles suggesting that they should reduce their consumption of health care materials and services? Two shifts in standard moral concepts with which people are commonly educated might help here. First, many environmental philosophers work from a concept of personal identity that appreciates individuals as strongly connected with all humans, creatures, and the natural world in a cyclical flow of materials and energy.[54,55] The concept of "ecosystem health" draws on the close relationship of human health to the condition of nature and makes it conceptually immediate to think of human health as dependent on ecosystem health.[56] This more holistic self-concept can help individuals to accept an extended sense of responsibility through an appreciation of connectedness with others and the natural world.

Second, part of our sense of personal identity and integrity depends on our ability to assume responsibilities to others, and so a moral conversation conscious of the need for people to meet their responsibilities could help to fulfill more completely the humanness of individuals. This Kantian approach ultimately rests the freedom of the individual on his or her ability to fulfill his or her sense of duty.[10,57,58] In health care ethics, it would thus become acceptable to expand the ethics conversation over "What does this patient want?" to include "How can we help this patient fulfill his or her sense of responsibility?" Principles of environmental responsibility and awareness of environmental effects need to be built into health care education and decisions at every level.

Key Ethical Tensions

To establish an ethical balance between environmental concerns and a commitment to patient care, three major dilemmas must be addressed: the individual versus the whole, sustainability versus social justice, and sustainability versus health.

The Individual versus the Whole

The Hippocratic principle of "do no harm" has strongly influenced the ethics of health care. This principle also requires that health care maintain sustainable practices and avoid harm to both humans and the natural world. Although medicine is already rife with dilemmas in which avoiding harm to patients results in harm or burdens to others, broadening our ethical responsibilities to include nature and future generations will undoubtedly intensify such conflicts. Perhaps the most difficult will be the inevitable conflicts between the individual and the whole. In medical ethics, the tendency has been to emphasize the responsibility of the health care provider to the individual patient—the relationship of trust; the need for the physician to keep the patient's benefit foremost. Global environmental concerns press physicians and other health care professionals to ask first "How will this commitment to care for patients place long-term burdens on the sustainability of health care?"

Sustainability versus Social Justice

Environmental sustainability and social justice are mutually reinforcing goals, and both are vital elements of population health. Yet the easy alliance of these two extraordinarily idealistic goals arouses a deeper sense of unease. The scale of these questions is so broad, and the empirical data available on their interrelationships so minimal, that one can only speculate.[59] Can industrialized countries in the Northern Hemisphere support their high levels of health care consumption without exploiting or ignoring widespread poverty, environmental degradation, ill health, and suffering in poorer regions of the world? The scope of the world's present distributive injustice—and the sheer number of people struggling to live with almost nothing—coupled with the profound constraints of our already-stressed ecosystems call into question our ability to achieve both sustainability and justice. We may have to ask which should have primacy.

Sustainability versus Health

The need for limits suggests potential problems for maintaining good health conditions in the long term. As the 1993 World Bank Development Report showed, health improvement in the twentieth century was closely linked to economic development.[53] Improved health and life expectancy were afforded by industrial and technological growth that stabilized food supplies, processed sewage, cleaned and transported water, developed vaccines, improved education, established health records and surveillance, and devised effective medical technologies.[60]

However, seeking gains in human health and welfare through aggressive economic development without regard for environmental effects may guarantee the ecological disaster already at our doorstep. Indeed, the increasing intensity of the agricultural, industrial, and energy sectors of the economy can be connected with increasing public health problems.[61] And if the world's most developed economies reduce their overall consumption of natural resources and materials to achieve sustainability, and if their health service industries thereby accomplish parallel reductions, will these health services become less effective? There is some evidence from less developed nations that good public health can be maintained on minimal resources when these resources are appropriately directed at basic public health infrastructures such as clean air and water, sanitation, education, and stable food supplies.[62,63] Will hospitals and clinics in the most highly technological and developed economies be able to learn to treat patients effectively while using fewer natural resources? If a significant scaling down of the health care sector is necessary, this may mean that some acutely ill individuals needing costly therapies will not receive treatment.[44] Will the already troubling ethical issues of rationing high-tech care necessarily be extended to an even wider range of health care services?[64]

Prescriptions

Health care professionals can offer leadership both in devising environmentally sound health care practices and in articulating the principles of sustainable health. Jane Lubchenco has urged scientists to undertake a new social contract to redirect the research enterprise from immediate social benefits toward a sustainable biosphere.[65] Similarly, health care professionals need to include environmental care among their primary ethical obligations.[2] This obligation can first be expressed by increasing consciousness of environmental issues in the education of health care professionals and patients, but it must eventually lead to appreciable reductions in the material scale of the world's most advanced health care services. Health care professionals will thus have to become actively involved in the ethical debates concerning balancing environmentally responsible health care with clinical services.

If the bright vision of the World Health Organization's fiftieth year is to be sustained, bioethics and health policy must begin to speak with one voice both to the needs of individuals and to the limitations of nature. Because earth's biological systems are necessary for human well-being, our medical pursuit of health must not diminish the abundance and vitality of the natural world.

Acknowledgment

Work for this chapter was supported by funds from The Greenwall Foundation and the Richard and Rhoda Goldman Fund.

References

1. Leaf A. Potential health effects of global climatic and environmental changes. *N Engl J Med* 1989; 321:1577–1583.

2. McCally M, Cassel CK. Medical responsibility and global environmental change. *Ann Intern Med* 1990; 113:467–473.

3. McMichael AJ. *Planetary overload: Global environmental change and the health of the human species.* Cambridge, MA: Cambridge University Press, 1993.

4. Last J. Human-induced ecological determinants of infectious disease. *Ecosystem Health* 1998; 4:83–91.

5. McCally M. Environment and health: An overview. *CMAJ* 2000; 163:533–535.

6. World Health Organization. The *world health report 1998. Life in the 21st century: a vision for all.* Geneva: World Health Organization, 1998.

7. Ewert AW, Kessler WB. Human health and natural ecosystems impacts and linkages. *Ecosystem Health* 1996; 2:271–278.

8. Rapport DJ. The dependency of human health on ecosystem health. *Ecosystem Health* 1997; 3:195–196.

9. Karr JA. Bridging the gap between human and ecological health. *Ecosystem Health* 1997; 3:197–199.

10. Westra L. *Living in integrity: A global ethic to restore a fragmented earth.* Lanham, MD: Rowman & Littlefield, 1998.

11. Soskolne CL, Bertollini R. *Global ecological integrity and "sustainable development" cornerstones of public health; December 3, 1998–December 4, 1998.* Rome: World Health Organization European Centre for Environment and Health, 1999.

12. Chen LC. World population and health. In: *Institute of Medicine. 2020 vision: Health in the 21st century.* Washington, DC: National Academy Press, 1996. pp. 15–23.

13. Daily G, editor. *Nature's services: Societal dependence on natural ecosystems.* Washington, DC: Island Press, 1997.

14. US Environmental Protection Agency. *Medical waste incinerators—Background information for proposed standards and guidelines; industry profile report for new and existing facilities.* Research Triangle Park, NC: Office of Air Quality Planning and Standards, US Environmental Protection Agency, 1994.

15. Davies T, Lowe AI. *Environmental implications of the health care service sector.* Washington, DC: Resources for the Future, 1999.

16. Garvin ML. *Infectious waste management: A practical guide.* Boca Raton, FL: Lewis Publishers, 1995.

17. Reinhardt PA, Gordon JG. *Infectious and medical waste management.* Chelsea, MI: Lewis Publishers, 1991.

18. Turnberg WL. *Biohazardous waste risk assessment, policy, and management.* New York: Wiley, 1996.

19. Hettenbach T. *"Greening" hospitals: An analysis of pollution prevention in America's top hospitals.* Falls Church, VA: Environmental Working Group/The Tides Center and Health Care Without Harm, 1998.

20. Tickner J. *The use of di–2-ethylhexyl phthalate in PVC medical devices: Exposure, toxicity, and alternatives.* Lowell, MA: Lowell Center for Sustainable Production, University of Massachusetts–Lowell, 1999.

21. Tickner JA, Schettler T, Guidotti T, McCally M, Rossi M. Health risks posed by use of di–2-ethylhexyl phthalate (DEHP) in PVC medical devices: A critical review. *Am J Industrial Med* 2001; 39:100–111.

22. Sustainable hospitals Project. Lowell, MA: Lowell Center for Sustainable Production, University of Massachusetts–Lowell. Available: www.uml.edu/centers/LCSP

23. Jameton A, Pierce J, Bowman A, Kerby C. *The Green Health Center.* Omaha: University of Nebraska Medical Center. Available: www.unmc.edu/green (accessed 2000 Dec 19).

24. Pierce J, Kerby C. The global ethics of latex gloves: Reflections on natural resource use in healthcare. *Camb Q Healthc Ethics* 1999; 8(1):98–907.

25. Wackernagel M, Rees WE. *Our ecological footprint: Reducing human impact on the earth.* Gabriola Island, BC: New Society Publishers, 1996.

26. Meadows DH, Meadows D, Randers J. *Beyond the limits: confronting global collapse; envisioning a sustainable future.* Post Mills, VT: Chelsea Green, 1992.

27. Cohen JE. *How many people can the earth support?* New York: WW Norton, 1995.

28. Smail JK. Beyond population stabilization: The case for dramatically reducing global human numbers. *Polit Life Sci* 1997; 16:183–192.

29. Durning A. How much is enough? *The consumer society and the future of the Earth.* Worldwatch Environmental Alert Series. New York: WW Norton, 1992.

30. Wackernagel M, Onisto L, Linares AC, Falfán ISL, Garcıía JM, Guerrero AIS, et al. *Ecological footprints of nations: How much nature do they use? How much nature do they have?* Xalapa: Mexico Centre for Sustainability Studies, Universidad Anáhuac de Xalapa, 1997.

31. Athanasiou T. *Divided planet: The ecology of rich and poor.* Boston: Little, Brown, 1996.

32. Brown LR. Nature's limits. In: Brown LR. *State of the world 1995: A Worldwatch Institute report on progress toward a sustainable society.* New York: WW Norton, 1995. pp. 3–20.

33. Butler CD. Overpopulation, overconsumption, and economics. *Lancet* 1994; 343:582–584.

34. Ekins P. The sustainable consumer society: A contradiction in terms? *Int Environ Affairs* 1991; 3:243–258.

35. Goodland R, Daly HE. Ten reasons why Northern income growth is not the solution to Southern poverty. In: Goodland R, Daly HE, Serafy SE, editors, *Population, technology, and lifestyle*. Washington, DC: Island Press, 1992. pp. 128–145.

36. Jameton A, Pierce JM. Toward a sustainable US health policy: Local congruities and global incongruities. *Soc Indicators Res* 1997; 4:125–146.

37. Reijnders L. The factor X debate: Setting targets for eco-efficiency. *J Ind Ecology* 1998; 2:13–22.

38. Smil V. *Energy in world history*. Boulder, CO: Westview Press, 1994.

39. Peet J. *Energy and the ecological economics of sustainability*. Washington, DC: Island Press, 1992.

40. Beauchamp TL, Childress JF. *Principles of biomedical ethics*. 4th ed. New York: Oxford University Press, 1994.

41. Jonsen AR. *The birth of bioethics*. New York: Oxford University Press, 1998.

42. Jonsen AR, Siegler M, Winslade WJ. *Clinical ethics*. 4th ed. New York: McGraw-Hill, 1998.

43. Wogaman JP, editor. *The population crisis and moral responsibility*. Washington, DC: Public Affairs Press, 1973.

44. Callahan D. *False hopes: Why America's quest for perfect health is a recipe for failure*. New York: Simon & Schuster, 1998.

45. Callahan D. *What kind of life: The limits of medical progress*. New York: Simon & Schuster, 1990.

46. Battin MP. *Sex and consequences: World population growth vs. reproductive rights*. Salt Lake City: University of Utah Press, 1994.

47. Partridge E, editor. *Responsibilities to future generations: Environmental ethics*. Buffalo, NY: Prometheus Books, 1981.

48. Weiss EB. Intergenerational equity toward an international legal framework. In: Choucri N, editor. *Global accord: Environmental challenges and international responses*. Cambridge, MA: MIT Press, 1993. pp. 333–354.

49. Norton BG. *Toward unity among environmentalists*. New York: Oxford University Press, 1991.

50. Loh J, Randers J, MacGillivray A, Kapos V, Jenkins M, Groombridge B, et al. *Living Planet Report 1998*. Gland, Switzerland: World Wide Fund for Nature, 1998.

51. Doll R. Health and the environment in the 1990s. *Am J Public Health* 1992; 82:933–941.

52. Scheper-Hughes N. *Death without weeping*. Berkeley: University of California Press, 1992.

53. World Bank. *World development report 1993: Investing in health.* Oxford: Oxford University Press, 1993.

54. Todd NJ, Todd J. *From eco-cities to living machines: Principles of ecological design.* Berkeley, CA: North Atlantic Books, 1994.

55. Wilson EO. *Consilience: The unity of knowledge.* New York: Knopf, 1998.

56. Rapport DJ, Böhm G, Buckingham D, Cairns J Jr, Costanza R, Karr JR, et al. Ecosystem health: The concept, the ISEH, and the important tasks ahead. *Ecosystem Health* 1999; 5:82–90.

57. Kant I [Gregor M, editor and translator]. *Groundwork of the metaphysics of morals.* New York: Cambridge University Press, 1998.

58. Joint Commission on Accreditation of Healthcare Organizations. Rights and responsibilities of patients. In: Joint Commission on Accreditation of Healthcare Organizations. *Accreditation manual for hospitals, 1990.* Chicago: Joint Commission on Accreditation of Healthcare Organizations, 1989. pp. xiii–xvii.

59. Dobson A. *Justice and the environment: Conceptions of environmental sustainability and dimensions of social justice.* Oxford: Oxford University Press, 1998.

60. McKeown T. *The role of medicine: Dream, mirage, or nemesis?* Princeton, NJ: Princeton University Press, 1979.

61. United Nations Development Programme, United Nations Environment Programme, World Bank, World Resources Institute. *World resources 2000–2001: People and ecosystems—the fraying web of life.* Amsterdam: Elsevier Science, 2000.

62. Caldwell JC. Routes to low mortality in poor countries. *Popul Devel Rev* 1986; 12:171–220.

63. Halstead SB, Walsh JA, Warren KS, editors. *Good health at low cost.* New York: Rockefeller Foundation, 1985.

64. Ubel PA. *Pricing life: Why it's time for health care rationing.* Cambridge, MA: MIT Press, 2000.

65. Lubchenco J. Entering the century of the environment: A new social contract for science. *Science* 1998; 279:491–497.

Index

Acetylcholine, 154
Acetylcholinesterase, 154
Acheson-Lilienthal report, 225
Acid rain, 17, 20
Acids, 17
Adaptation, precautionary principle and, 252
Adenoviruses, 46
Adipose tissue, industrial chemicals in, 167–173, 176–177
Administrative controls, for occupational disease, 265
Adverse health effects. *See also under specific pollutants*
  of air pollutants, 18
  of background exposure, 186–189
  definition of, 17
  exposure and, 233
  frequency/magnitude of, 233
  individuality of, 233
Advocacy organizations, 5
Aerial bombardment, 275–276
Aerosols, acidic, 19–21
Africa
  AIDS epidemic, 88
  insect-borne diseases in, 103
  land mines in, 276
  population growth in, 87–88
  pregnancy-related deaths in, 90
Age
  body-burden variation and, 182
  at breast development, endocrine disruption and, 152
  heart-rate variability-air pollution relationship and, 23
Agency for Toxic Substances and Disease Registry (ATSDR), 50, 65, 267
Agriculture
  climate change and, 102, 109–110
  corporatization of, 4
  pesticide usage (*see* Pesticides)
  water withdrawals for, 40–41
AIDS, 88, 90
Air pollution. *See also specific air pollutants*
  adverse respiratory effects, 18–29
  biological impact of, 105
  carcinogens in, 28
  crisis, 16
  developing countries and, 16
  indoor, 29–30
  industrial chemicals and, 163
  industrialized countries and, 16
  particulate (*see* Particulate matter)
  regulations, current, 30–32
  respiratory disease and, 15
  urban, 16–17, 19
Algae, 109

Allergens
adverse health effects, 18
sources, 18
Alpha particles, 213
Aluminum, 76
Alzheimer's disease, aluminum exposure and, 76
American Cancer Society, 207
American Public Health Association, 179
Animals. *See* Species loss; Species preservation
Animal studies
advantages of, 8
disadvantages of, 9
extrapolating to humans, 9, 236
nitrogen dioxide, 25
ozone, 25
Antarctica, 111, 137
Antibiotics, 121
Arctic region, 111
Argentina, 51
Arsenic
adverse health effects, 18, 51, 74
exposure, 73
forms of, 73
sources, 18, 51
toxicity, 73–74
water contamination, 51–52
Asbestos
adverse health effects, 18
mesothelioma and, 262
sources, 18
and tobacco smoke, 204–205
Asia
insect-borne disease in, 103
population growth, 87–88
pregnancy-related deaths, 90
Southeast, sea-level rise and, 110
Aspirin, 121
Asthma, 19, 20, 258
Atomic bomb survivors, 214, 216–217
Atrazine, 49, 150
ATSDR (Agency for Toxic Substances and Disease Registry), 50, 65, 267
Autonomy, 288–289
Bacterial pathogens, waterborne, 45
Bangladesh, 51–52, 109
Basal cell carcinoma, 137–138
Bears, 123–124
Benzene, 164
Beta particles, 213
Biodiversity
alterations, 125
in emergence/spread of disease, 128
global assessment of, 127
microbial and marine, 128
threats to, 120
Biological markers of exposure, 265
Biomass, indoor air pollution and, 29–30
Biomonitoring of metals, population-based, 77
Biotoxins, 109
Bird species, 2
Birth-defect registries, 157
Birth rates, 84–86
Black bear *(Ursus americanus),* 123–124
Bladder cancer, 53, 236
Body burden
definition of, 163
measurement of, 164–166, 186
"normal background," 180–181, 186–189
public health and epidemiologic significance, 186–187
public understanding of, 190
variations in, 182–184
Botswana, 88
BRAVO test, 218
Brazil, 17, 48, 71, 104
Breast cancer, 152, 218
Breast milk
dioxin in, 185
industrial chemicals in, 167–173, 176
Building materials, indoor air pollution and, 29–30
Burden of proof, precautionary principle and, 251–252
Bush, George W., 12, 250

Cadmium, 2, 74–75
Caenorhabditis elegans, 125
California, 27, 48–49

Canada
cancer incidence, 201
consumption patterns, impact on environment, 87
population growth, 86
sea-level rise and, 110–111
vector-borne disease, 107–108
Canadian National Pollutant Release Inventory, 207
Cancer. *See also* Carcinogens; *specific carcinogens*
arsenic and, 74
bladder, 53, 236
breast, 152, 218
childhood, 258
clusters, 201
endocrine disruption and, 152
environmental causes of, 204–206
environmental exposure assessment, 207–208
estimated lifetime risk, 204
gastric, 51
incidence, 201
industrial discharges and, 50
ionizing radiation and, 215–219
low-level radiation and, 217
lung (*see* Lung cancer)
mortality rates, 201–203
non-Hodgkin's lymphoma, 140–141, 202, 205
occupational exposures and, 204–206
oral, 236
pesticides and, 49
prevention, 206–207
prostate, 152
radon exposure and, 50–51
rates, 208
risk assessment, 8
skin (*see* Skin cancer)
squalamine and, 124
testicular, 152
thyroid, ionizing radiation and, 217–220
trends, 201–204
Carbon dioxide emissions
increases in, 1, 2, 4
reduction of, 92, 111–112
Carbon monoxide (CO)
adverse health effects, 18, 27–28
affinity for hemoglobin, 27
chronic low-level exposure, 28
in developing countries, 16, 30
maximum allowable exposure, 21
poisoning, 27–28
sources, 18, 27
Carboxyhemoglobin (COHb), 27–28
Carcinogens. *See also specific carcinogens*
in air pollution, 28
occupational/environmental, 205
risk estimators for, 9
Cardiovascular disease, particulate levels and, 22–23
Cataracts, 4, 139–140
Catastrophic losses, weather-related, 103
Causation, correlation and, 246
CDC (Centers for Disease Control), 267
Cellular telephones, adverse health effects, 223, 224
Centers for Children's Environmental Health and Prevention Research, 266–267
Centers for Disease Control (CDC), 267
Cesium-137, 221
CFCs (chlorofluorocarbons), ozone depletion and, 136–137, 139, 244
Chemicals, industrial. *See also specific chemicals*
biomonitoring in U.S., 166–178
body burden (*see* Body burden)
children's exposure to, 184–185
discharges, 49–50
health risks from, 5
industrial discharges, 49–50
manufacturing of, 2
organochlorine (*see* Organochlorines)
pesticides (*see* Pesticides)
public health and epidemiologic significance, 186–187
as water quality threats, 47–53
women's exposure to, 184–185

Chernobyl nuclear accident, 211, 213, 215, 218–219, 224
Children
adverse health effects of background exposure, 186–189
background exposures in Great Lakes area, 188
case studies, 260
chemical exposures, 258–259
environmental health needs of, 266
illness patterns, 258
industrial chemical exposure, 184–185
lead exposure, 68–70, 76–77, 258, 260
leukemia in, 217, 219, 222, 260
mercury toxicity, 72
metabolic pathways, 259
mortality rates, reproductive health programs and, 90
PCB exposure, 153
protection from environmental toxins, 261
respiratory disease mortality and, 17
subclinical toxicity, 261
unique vulnerability of, 259–260
Children's Environmental Protection Act, 266
"Children's Framingham" study, 185
Chile, 51
China
air pollution in, 16
carbon dioxide emissions, 112
coal reserves, 19
economic development, 92
fertility rates, 85–86
indoor air pollution, 24
Chloramine, in water, 43
Chlorinated solvents, 50
Chlorination, protozoal disease and, 43, 46–47
Chlorofluorocarbons (CFCs), ozone depletion and, 136–137, 139, 244
Chloropyrifos, 184
Cholera transmission, 109
Chondrodendron tomentosum, 121
Chromium, 76
Chronic airway obstruction, sulfur dioxide and, 21
Cinchona officianalis, 121
CITES (Convention on International Trade in Endangered Species), 127
Clean Air Acts, 19, 30–31
Clean Water Act, 42, 49
Climate change
adaptation to, 112
El Niño and, 101
evidence of, 88
greenhouse effect and, 99, 100
greenhouse gases and, 99–101
health impacts of, 104–111
human impact on, 3–4
ocean warming, 101–102
population vulnerability and, 111
prescriptions for, 111–112
signals of, 102–104
Coal, 19
COHb (carboxyhemoglobin), 27–28
Coke ovens, lung cancer risk and, 28
Coliform bacteria tests, 44–45
Community exposures, cancer and, 201, 206
Cone snails, 122–123
Conotoxins, 122–123
Consumption, human, 86–89
Contraceptives, 91
Cooking fuels, indoor air pollution and, 29–30
Correlation, causation and, 246
Coxsackie B virus, 46
Cryptorchidism (undescended testicles), 151, 152
Cryptosporidiosis, 47, 109
Cryptosporidium parvum, 45, 46
Cytarabine, 121
Czech Republic, 17

2,4-D, 49
DBCP (dibromochloropropane), 48–49
DDE, 151, 183, 188
DDT
adverse health effects, 49, 148, 149, 151
in blood/urine, 155
body burdens, time trends in, 184
brain development and, 154
global phase out of, 157

mechanism of action, 149
in water, 48
Deaths. *See* Mortality rates
Defoliation, 278–279
Deforestation, 4
DEHP (di-2-ethylhexyl-phthalate), 150
Demographic factors, of global warming, 105
Dengue fever, 44, 103, 108
Denning, 123–124
Dental amalgams, mercury exposure from, 72
Depo-Provera, 91
DES (diethylstilbestrol), 148, 149
Despoliation, 278–279
Developing countries. *See also specific developing countries*
air pollution and, 16–17
carbon monoxide exposure, 16, 30
deaths in, 88
family planning programs in, 89–91
family size in, 89
fertility rates, 85
indoor air pollution and, 29–30
industrialized countries and, 11–12
lead exposure in, 68
nitrate exposure and, 47–48
particulate levels in, 23, 24
pesticide exposure in, 48
population growth in, 89–90
reproductive health programs, 90–91
waterborne disease in, 45
Development projects, environmentally unsound, 92
Diabetes mellitus, 46
Diarrhea, 39, 46
Dibromochloropropane (DBCP), 48–49
Dieldrin, 48, 49, 151, 184
Di-2-ethylhexyl-phthalate (DEHP), 150
Diethylstilbestrol (DES), 148, 149
Digitalis, 121
1,25-Dihydroxyvitamin D, 137
Dimethylmercury, 72
Dinoflagellates, toxic, 109
Dioxin
adverse health effects, 150
body burden, 186–187
in breast milk, 185
cancer implications, 206–207
exposure, 180, 181, 201
low-level exposures, 155
mechanism of action, 150
sources, 5, 206–207
Disinfection by-products, 52–53
Dose, absorbed, 7
Dose-response assessment, 7
Dose-response modeling, 232
Dracunculiasis, 44
Drinking water. *See* Water
D-tubocurarine, 121

Earth surface temperatures, changes in, 99–102
Ecological Footprints of Nations Study, 287–288
Economic development, 91–93, 105
Ecosystem
definition of, 125
disturbances from global warming, 105
health of, 289–290
services, 125–126, 127–128
Ecosystem-based health perspective, 3–4
Education
of health care professionals, 10–11
lack of, 89
long-term strategies, 12
on metal toxicity, 77
National Fellowship Training Program for, 267
underdiagnosis of occupational disease and, 263–264
Egypt, 55
Electromagnetic radiation
adverse health effects, 222–223
exposure, limitation/prevention strategies, 224
fields, 213–214
spectrum of, 211, 212, 213
El Niño, 101, 104

Encephalitis, 103, 107
Encephalopathy, solvent-induced, 262
Endocrine disruptors. *See also specific endocrine disruptors*
epidemiologic studies, 151
global phase out of, 157
health implications, 149–151
historical background, 147
human disease trends and, 151, 152
low-level exposures, 154–155
mechanism of action, 148–149
neurobehavioral effects, 151, 153
policy implications, 154–155
reproductive effects, 151
screening, testing and tracking, 157–158
Engineering controls, for occupational disease, 264
ENSO cycle (El Niño Southern Oscillation Cycle), 101, 106, 109
Enteroviruses, 46
Environmental degradation
cancer etiology and, 204–206
crisis of, 1–2
global, 4–6
health and, 4–6
population, consumption, health concerns and, 86–89
poverty, population growth and, 4–5
public concerns, 5–6
stopping, 11–12
Environmental education. *See* Education
Environmental exposure assessment, cancer and, 205, 207–208
Environmental health policies
alternative assessment, 248–249
goal setting, 248
Environmental pediatrics education, 267
Environmental Protection Agency (EPA)
air quality standards, 19, 20–21, 30
burden of proof and, 251–252
industrial chemicals and, 49, 163
lead exposure and, 52
medical-waste incineration, 206–207
National Human Adipose Tissue Survey, 176–177
Office of Children's Health Protection, 267
Organization for Economic Cooperation and Development, 157
TEAM study, 177
Toxics Release Inventory, 207
trihalomethanes and, 53
Environmental sustainability, *vs.* health, 291–292
EPA. *See* Environmental Protection Agency
Erethism, 72
Escherichia coli, 45
Ethics
"do no harm," 290
environmental-patient care balance and, 290
individual *vs.* societal concerns and, 290
Europe, 86, 87. *See also specific European countries*
Exposure
adverse responses and, 233
assessment, 7, 232
definition of, 6–7
"normal background," 180–181, 186–189
reference ranges, 178
timing of, 9
tracking, 267
Exposure-disease model, 7

False negative (type II error), 250–251
False positive (type I error), 251
Family planning programs
cost of, 91
in developing countries, 89–90
Fetus
ionizing radiation exposure, 260
lead exposure, 70
mercury exposure, 72
vulnerability, 148–149
FEV_1 (forced expiratory volume in 1 second), 26

Fish
cichlid, overfishing of, 126
methylmercury contamination, 71, 260
Food and Agriculture Organization (FAO), 128
Food-borne disease, 105, 106
Food chain, organochlorine-contaminated, 153
Food poisoning, 104
Food production
climate change and, 109–110
global warming and, 105
for world population, 6
Forced expiratory volume in 1 second (FEV_1), 26
Forestry, corporatization of, 4
Formaldehyde, 164
Fort Dix, New Jersey, 26
Fossil-fuel combustion products, 19, 112. *See also* Sulfur dioxide
Foxglove plant, 121
Framingham Heart Study, 23
Frogs, 2
Fundamental uncertainty, precautionary principle and, 245–246
Furans, 186

GAC (granulated activated charcoal), 55
Gambia, 24
Gamma rays, 213
Gas chromatography/mass spectrometry (GC/MS), 164, 165
Gasoline, leaded, 77
Gastric cancer, 51
Gastrointestinal infections, 104
GC/MS (gas chromatography/mass spectrometry), 164, 165
Gender, body-burden variation and, 182
Geography, body-burden variation and, 182–183
Germany, 224
Giardia lamblia, 45, 46
Global Environmental Facility, 127
Global warming. *See* Climate change
Goal setting, for environmental health policies, 248
Governmental agencies. *See also specific governmental agencies*
risk assessment and, 237–238
Granulated activated charcoal (GAC), 55
Great Britain, 21, 22, 30
Great Lakes region
International Joint Commission and, 155–157
toxic contamination of food chain in, 153, 188, 240
Greenhouse effect, 99, 100
Greenhouse-gas emissions. *See also* Carbon dioxide emissions; Nitrogen oxides
earth surface temperatures and, 99–102
industrial growth and, 99, 101, 112
limited reductions, 92, 112
stabilization of, 111
sulfur oxide, 16, 18
Greenland, 101–102, 110
Groundwater, 40

Haloacetic acids (HAAs), 53
Hanford Environmental Dose Reconstruction Project, 220
Hantavirus pulmonary syndrome, 109, 126
Harm, potential for, 241, 243
Harvard Six Cities Study, 19–20, 22
Hazard identification, 7, 231–232
Health
environment and, 4–6, 86–89
occupational, future needs for, 266–268
population and, 86–89
problems, environmental, 1
vs. environmental sustainability, 291–292
water scarcity and, 88
Health care
as environmental problem, 286, 287
as environmental solution, 286–287
environmental sound practices, 292
sustainable, 287–290
Health care industry, 5
Health care professionals, environmental education of, 10–11
Heart-rate variability (HRV), 22–23

Heating fuels, indoor air pollution and, 29–30
Heavy metals. *See* Metals
Hemoglobin, 27
Hepatitis A virus, 46
Heptachlor, 184
Herbicides, 48, 49
Herpes simplex cold sore, 140
Hill criteria, 246
Homo sapiens. *See* Humans
Hormonally active agents. *See* Endocrine disruptors
Hormones
  low-level exposures, 154
  neurobehavioral effects, 151, 153
"Hot spots," ecological, 92, 128
Household appliance radiation, adverse health effects from, 223
HPG (hypothalamic-pituitary-gonadal axis), 153
HRV (heart-rate variability), 22–23
Human activities
  ecosystem disruption by, 119
  environmental crisis and, 1–2
  environmental impact of, 239
  impact on climate, 3–4
  species extinction and, 119–120
Humans
  adaptation to UV stress, 135–136, 137
  body fat, industrial chemicals in, 167–173, 176–177
  breast milk, industrial chemicals in, 167–173, 176
  ecosystem-based health perspective and, 3–4
Hurricane Mitch, 104–105
Hydrocarbons, 5, 181
Hypospadias, 151, 152
Hypothalamic-pituitary-gonadal axis (HPG), 153
Hypothyroidism, 148–149

IARC (International Agency for Research on Cancer), 206
II-131, 224
IJC (International Joint Commission), 155–157, 240
Illness
  from extreme environmental effects, 104
  occupationally-related. (*See* Occupational disease)
Immigration, 86
Immunosuppression, ultraviolet radiation-induced, 140–141
India
  arsenic levels in, 51–5
  carbon dioxide emissions, 112
  Delhi, 16–17
  indoor air pollution, 24
  nitrate exposure and, 48
  nuclear weapons and, 225
  population growth, 88
  water resource conflicts, 55
Indonesia, 104
Industrial chemicals. *See* Chemicals, industrial; *specific chemicals*
Industrialization rate, 2
Industrialized countries
  adoption of transparent, inclusive open processes, 249–250
  air pollution and, 16
  developing countries and, 11–12
  nitrate exposure and, 47–48
  water withdrawals, 40–41
Infant mortality, 17, 90, 258
Information, on war-related environmental destruction, 279
Injury, from extreme environmental effects, 104
Insect-borne disease, global warming and, 103
Insecticides, 48, 154. *See also* Organochlorines; *specific insecticides*
Insects, global warming and, 103
Intergovernment Panel on Climate Change (IPCC), 3–4, 100, 127
Internal combustion engine emissions, 27
International Agency for Research on Cancer (IARC), 206
International Commission on Radiological Protection, 216
International Joint Commission (IJC), 155–157, 240

International law, 280
Intestinal infections, 104
Ion exchange resins, 55
Ionizing radiation
adverse health effects, 214–216
exposure dosage nomenclature, 214–215
exposure standards, 216
fallout from nuclear testing, 218
features of, 213
forms of, 213
high-level doses, 214–215
human exposure to, 211, 212
limitation/prevention strategies, 223–224
linear, no-threshold hypothesis, 215
low-level doses, 215
from nuclear weapons production/testing, 274–275
IPCC (Intergovernmental Panel on Climate Change), 3–4, 100, 127
IQ, lead exposure and, 70
Irai-itai disease, 75
Iraq, 55, 225
Iraq-Iran war, 278
Irrigation, 40–41, 57
Israel, 55
Ixodes ricinus, 103

Japan
aerial and naval bombardment, 275–276
consumption patterns, impact on environment, 87
mercury contamination, 260
population decline, 86
Jordan, 55

Kenya, 24
Kidney toxicity, of cadmium, 75
Kosovo, land mines in, 276
Kyoto Climate Change Convention, 112, 250

Latin America. *See also specific Latin American countries*
insect-borne disease in, 103

Lead exposure
adverse health effects, 52, 68–71
blood levels, 70, 176
in children, 258, 260
control, under current U.S. standards, 31
in developing countries, 68
low-level, 65
maximum allowable, 21
modifiers of, 68
reducing, 76–78
routes, 65–66
sources, 2, 52, 67–68
subclinical toxicity in children, 261
toxicity of, 68–71
uses of, 67–68
water contamination and, 52
Learning, precautionary principle and, 252
Legionella, 44
Leprosy, 140
Leukemia, childhood, 217, 219, 222, 260
Libya, 276
Linear, no-threshold hypothesis, 215
Lowest dose observed effect level (LOEL), 235
Lung cancer
mesothelioma, 262
occupational risk, 28
radon-related, 51, 221–222
work-related, underdiagnosis of, 262
Lyme disease, 103, 109, 125

Malaria, 44, 103, 107, 108, 140
Manganese, 75–76
Margin-of-safety approach, 234–235
Marshall Island, 218
Mathematical models, of climate change effects on vector-borne disease, 108
MEA (Millennium Ecosystem Assessment), 127
Measles, 88
Medical education. *See* Education
Medical-waste incineration, 5
Medicines. *See also specific medications*
potential, 120–121, 128

Melanoma, 205
Melatonin hypothesis, 213–214
Mercury
exposure, 65, 71
forms of, 71
from medical-waste incineration, 5
toxicity, 72–73
Mesothelioma, 262
Metabolic models, 164–165
Metals. *See also specific metals*
adverse health effects, 66
exposure routes, 65–66
monitoring pollution worldwide, 77
production moratorium, 78
reducing exposure to, 76–78
toxicity of, 66
Methane, 2
Methemoglobinemia, 47
Methoxychlor, 149
Methylcyclopentadienyl manganese tricarbonyl (MMT), 75–76, 78
Methylmercury, 260
Methyl-tertiary-butyl-ether (MTBE), 49–50
Mexico, 17, 25
Millennium Ecosystem Assessment (MEA), 127
MMT (methylcyclopentadienyl manganese tricarbonyl), 75–76, 78
Mobile telephones, adverse health effects of, 223, 224
Model uncertainty, precautionary principle and, 245
Montreal, Canada, 106–107
Montreal Protocol of 1987, 139
Mortality rates
environmental temperature and, 106–107
from extreme environmental effects, 104
heat-related, 106–107
infant, 17, 90, 258
from nitrate poisoning, 47
population growth and, 84
from pregnancy-related causes, 90
from respiratory disease, 17
winter, 107
MTBE (methyl-tertiary-butyl-ether), 49–50
Mycobacterium avium species, 44

National Academy of Science (NAS)
Biological Effects of Ionizing Radiation report, 215, 216
blood sample monitoring, 177
greenhouse gases and, 3–4
hormonally active agents and, 180
risk assessment, 231
National Council on Radiation Protection and Measurement, 216
National Fellowship Training Program, in environmental pediatrics, 267
National Health and Nutrition Examination Survey (NHANES), 176, 177, 182, 183
National Human Adipose Tissue Survey (NHATS), 176–177, 182, 183, 185
National Institute For Occupational Safety and Health (NIOSH), 262
National Longitudinal Study, 266
National Report on Human Exposure to Environmental Chemicals program, 177–178
National Research Council, 165, 231
The Natural Step, 249
Naval bombardment, 275–276
Nematocide contamination, 48–49
Nepal, 17, 24
Neurobehavioral effects, of hormones, 151, 153
New Guinea, 24
New York, 20, 56
NHANES (National Health and Nutrition Examination Survey), 176, 177, 182, 183
NHATS (National Human Adipose Tissue Survey), 176–177, 182, 183, 185
Nigeria, 88
NIOSH (National Institute For Occupational Safety and Health), 262
Nitrates, 47–48
Nitrogen, 1–2

Nitrogen dioxide (NO_2)
adverse respiratory effects, 24–25, 26
animal studies, 24, 25, 26
in developing countries, 17
maximum allowable exposure, 21
Nitrogen oxides. *See also* Nitrogen dioxide
adverse health effects, 18
in smog, 107
sources, 16, 18, 24–25
NO_2. *See* Nitrogen dioxide
NOEL (no observed effect level), 235
Non-Hodgkin's lymphoma, 140–141, 202, 205
Nonionizing radiation
adverse health effects, 222–223
limitation/prevention strategies for, 223–224
radio-frequency range, 213
Non-point sources, 42
No observed effect level (NOEL), 235
North Korea, 225
Norwalk viruses, 45, 46
Nuclear-power plants
accidents, 211, 213, 218–219
environmental contamination from, 219–220
repository sites, 225–226
routine operations, 219
waste contamination, 220–221
Nuclear technology risks, 213
Nuclear weapons, 225, 274–275
Null hypothesis, 250–251

O_3. *See* Ozone
Occupational disease
diagnosis of, 262–264
needed federal programs for, 266–268
prevention of, 264–265
tracking, 267
types of, 262, 263
Occupational exposure. *See also* Occupational disease
arsenic, 73
body-burden variation and, 183–184
cancer risk and, 204–206
carcinogens, 205
electromagnetic radiation, 223
to endocrine disruptors, 151
latency period, 263
mercury, 72
Occupational health needs, future, 266–268
OCHP (Office of Children's Health Protection), 267
OECD (Organization for Economic Cooperation and Development), 157
Office of Children's Health Protection (OCHP), 267
Ohio, 21–22, 49
Oral cancer, 236
Oral contraceptives, 91
Oral rehydration therapy, 39
Organization for Economic Cooperation and Development (OECD), 157
Organochlorines, 156. *See also* Dioxin; Polychlorinated biphenyls
adverse health effects, 49, 179–180
body burden of, 180–181
lactational exposure, 188
lipophilicity of, 179
in North American population, 167–173
persistence of, 179
phase out of, 156
production, 178–179
Osteoporosis, animal research, 124
Ozone (O3)
adverse health effects, 18, 24–26, 137–141
air pollution, secondary, 17
animal studies, 25
control, under current U.S. standards, 31
depletion, 2, 136–137
diurnal variations, 25
federal standards, 25
high concentrations, 25
"hole," 5, 137
indirect health effects, 141
indoor *vs.* outdoor concentrations, 26
layer formation, 135, 136
maximum allowable exposure, 21
sources, 18, 24–25
thickness of layer, 139

Ozone-depleting substances. *See also specific ozone-depleting substances*
phasing out of, 141–142

Pacific Yew tree *(Taxus brevifolia),* 121–122
PAHs (polyaromatic hydrocarbons), 18
Pakistan, 55, 88, 225
Parkinson's disease, 76
Particulate matter (PM)
adverse health effects, 18, 28
from coal combustion, 22
definition of, 21
in developing countries, 23, 24
heart-rate variability and, 22–23
maximum allowable exposure, 21
from mobile sources, 22
particle size, 22
respiratory mortality and, 21–22
respiratory symptoms and, 22
sources, 18
PCBs. *See* Polychlorinated biphenyls
PCDFs (polychlorinated dibenzofurans), 167, 182
Peak expiratory flow, ozone and, 25
Pediatric environmental health specialty units (PEHSUs), 267
Perchloroethylene, 50
Persistent organic pollutants (POPs), 5, 157, 179. *See also specific persistent organic pollutants*
Personal protective equipment, 265
Pesticides. *See also* Organochlorines; *specific pesticides*
adverse health effects, 49
blood levels, 176
body burden, 164, 181, 184
contamination events, 48–49
exposure in critical populations, 184–185
mechanisms of action, 148
organophosphate, 154
sources, 48
Photochemical oxidants, 24. *See also* Nitrogen oxides; Ozone
Photodermatoses, 137
Physicians
environmental education of, 10–11
in occupational disease, 265
Pineal gland, 214
Plants. *See also* Species loss; Species preservation
migration of, 102–103
as potential medicines, 120–121
Plutonium, 65, 221
Plutonium-reprocessing industry, 225
PM. *See* Particulate matter
Point sources, 42
Polar bears *(Ursus maritimus),* 123–124
Polyaromatic hydrocarbons (PAHs), 18
Polychlorinated biphenyls (PCBs)
adverse health effects, 148, 149, 151, 153, 155, 188–189
body burdens, time trends in, 184
geographic variations in, 183
mechanism of action, 149
phase out, 157
in tissues/body fluids, 170–173
Polychlorinated dibenzofurans (PCDFs), 167, 182
Polycyclic aromatic hydrocarbons (PAHs), 28
POPs (persistent organic pollutants), 5, 157, 179. *See also specific persistent organic pollutants*
Population growth
consumption and, 86–89
displacement, 105
environmental health concerns and, 86–89
food production and, 110
food supply and, 6
poverty, environmental degradation and, 4–5
reduction, 112
statistical trends, 83–86
threats from, 83
vulnerability, climate change and, 111
water availability and, 53
Poverty, 4–5, 89
Power lines, adverse health effects from, 222

Precautionary action
options for, 252–253
precautionary principle and, 241, 247
Precautionary principle, 237
elements, core, 241–247
historical aspects, 240–241
implementation, 248–253
risk assessment and, 253–254
treaties/conventions and, 240–241
Predator species, alteration of, 126
Pregnancy
electric blanket usage during, 223
ionizing radiation exposure during, 260
lead exposure during, 70
mercury exposure during, 71
in utero exposures, 184–185
Primary prevention, 264
Probability, 7
Programmed cell death, 125
Prolactin, 151
Prostate cancer, 152
Protozoa, waterborne, 46–47
Protozoal disease, waterborne, 45
p53 tumor suppresser gene mutations, 137

Quinine, 121

Race, body-burden variation and, 183
Radiation. *See* Electromagnetic radiation; Ionizing radiation; Ultraviolet radiation
Radioactive fallout, 218, 224
Radioactive iodine, 206
Radioactive waste, 220–221
Radon
adverse health effects, 18, 50–51
indoor air concentrations, 50
inhalation, 221
lung cancer and, 221–222
progeny, 221
sources, 18, 50
and tobacco smoke exposure, 205
water contamination, 50–51
Reproductive effects
of endocrine disruptors, 151
of nitrate exposure, 48
Reproductive health programs, 90–91
Research
metal toxicity, 78
model loss, species loss and, 123
Respirable particulates, 16
Respiratory disease
acidic aerosols and, 19–20
air pollution and, 15
mortality, 17, 20–23
"Reverse onus" concept, 156
Reverse osmosis, 55
"Rice-water" stools, 45
"Right-to-Know," 266
Rio Declaration of 1992, 240–241
Risk. *See also* Risk assessment
acceptability of, 233–234
analysis, 6
characterization, 7
communication, 10
context of, 234–235
definition of, 6
direct measurement of, 233
identification, interdisciplinary, 231–232
management, 231, 248–249
statistics, 232–233
Risk assessment
cancer, 8
components of, 7
data for, 8
definition of, 6
exposures for, 234
governmental agencies and, 237–238
issues, 233
margin-of-safety approach, 234–235
precautionary principle and, 248–249, 253–254
refining, 9–10
reproductive, 9
scope of, 231–234
uncertainty in, 235–236
utility, 236–238
vulnerable populations, 257 (*see also specific vulnerable populations*)
Rodent populations, 109
Rotaviruses, 46

SAB (Science Advisory Board), 155–156
Saccharin, 236
Safe Drinking Water Act, 42, 49, 56
Safety factors, 235
Salix alba, 121
Salmonella, 45
Schistosomiasis, 44, 108, 126
Science Advisory Board (SAB), 155–156
Scientific uncertainty
precautionary principle and, 241, 243–244
scientific proof and, 246–247
Scientists, anticipation of environmental change, 5
Sea-level rise, 105, 106, 110–111
Sea surface temperature, 109
Secondary prevention, 264
Sex ratio, endocrine disruption and, 152
Sexually transmitted diseases (STDs), 90–91
Sharks, 124
Shigella, 45
Signaling pathways, 153–154
Simazine, 49
"Sixth extinction," 120
Skin cancer
immunosuppression and, 140
non-Hodgkin's lymphoma and, 140
ultraviolet light exposure and, 4, 137–138
Smallpox, 140
Smog, photochemical, 19, 135, 141
Smoking
oral cancer and, 236
radon-related lung cancer and, 221–222
SNX-111 (ziconotide), 122–123
Social factors, 89, 105
Social justice, *vs.* environmental sustainability, 291–292
Species loss
addressing causes of, 128
extinction, 119–120
from human activities, 239
predators, 126
protection efforts for, 126–129
research model loss and, 123
Species preservation, 119–120, 127
Sperm count, endocrine disruption and, 152
Squalamine, 124
Squamous cell carcinoma, 137–138
St. Louis encephalitis, 107, 108
State Watershed Assessment Plan (SWAP), 56
Statistical uncertainty
analysis of, 250–251
precautionary principle and, 244
STDs (sexually transmitted diseases), 90–91
Strontium-90, 221
Subclinical toxicity, in children, 261
Sudan, 55
Sulfur dioxide
adverse health effects, 19, 28
bronchoconstrictor effects, 20–21
in developing countries, 17
emissions, reduction of, 19
maximum allowable exposure, 21
sources, 19
Sulfur oxide, 16, 18
Superfund sites, 50
Sustainability, environmental, *vs.* social justice, 291–292
"Sustainable development" strategies, 12
SWAP (State Watershed Assessment Plan), 56
Sweden, 224
Synergistic exposure, 204–205

T3, 148
T4, 148
Taiwan, 51
Taxol, 121–122
Taxus brevifolia (Pacific Yew tree), 121–122
Tertiary prevention, 264
Testicles, undescended, 151, 152
Testicular cancer, 152
Testosterone, 151
Tethya crypta, 121
TFR (total fertility rate), 84–85, 90
Thailand, 91

Thermal extremes, exposure to, 105–107
THMs (trihalomethanes), 53
Threat assessment, for war-related technology, 279
Three Mile Island nuclear plant accident, 211, 213, 218
Thyroid cancer, ionizing radiation and, 217–220
Thyroid hormones, 148, 151, 153
Thyroid stimulating hormone (TSH), 151, 153
Ticks, 103, 109
Time-to-tumor models, 9
Time trends, body-burden variation and, 184
Tobacco smoke, 204–205
Tolerance distribution models, 9
Total fertility rate (TFR), 84–85, 90
Total suspended particulates (TSPs), 16–17
Trachoma, 44
Trematocranus plachydonan, 126
Trichloroethylene, 50
Trihalomethanes (THMs), 53
TSH (thyroid stimulating hormone), 151, 153
TSPs (total suspended particulates), 16–17
Tuberculosis, 88, 140
Tube wells, 51–52, 73
Tubocurarine, 121
Turkey, 55
Type I error, 251
Type II error, 250–251
Typhoid fever, 45

Ultraviolet radiation
  absorption by ozone layer, 135–136
  cataracts and, 139–140
  health threats from, 4
  immune effects, 140–141
  photochemical reactions, 24, 135
  skin cancer and, 137–138
Uncertainty, in risk assessment, 235–236
UNDP (United Nations Development Program), 127
UNEP (United Nations Environment Programme), 127, 128
United Nations. *See also specific United Nations programs*
  demographic projections, 84
  high-fertility model, 84
  Intergovernmental Panel of Climate Change, 3–4, 100, 127
  International Conference on Population and Development, 90, 91
  low-fertility model, 85
United Nations Development Program (UNDP), 127
United Nations Environment Programme (UNEP), 127, 128
United States. *See also specific governmental agencies; specific states*
  air pollution regulations, 30–32
  arsenic levels in, 51
  biomonitoring program, 166–178
  consumption patterns, impact on environment, 87
  emission controls, 27
  environmental legislation, 19, 30–31, 42, 49, 56
  limitations of air pollution regulations, 31
  population growth, 86
  population growth predictions, 84, 85
Uranium, 225
Ursus americanus (black bear), 123–124

Vaginal clear cell carcinoma, 148
Vector-borne disease
  in Canada, 107–108
  climate change and, 103, 104
  mathematical models, 108
Venezuelan equine encephalomyelitis, 107
Vibrio cholerae, 45, 109
Vietnam, 278
Vinblastine, 121
Vinca rosea, 121
Vinclozolin, 149
Vincristine, 121
Viral pathogens, waterborne, 46
Vitamin D, 139

Vitamin D3, 137
Vulnerability assessment, for war-related technology, 279–280
Vulnerable populations, children. *See* Children; Women

War, environmental destruction from, 273–274
  aerial and naval bombardment and, 275–276
  despoliation/defoliation and, 278–279
  land mines and, 276–278
  nuclear weapons production/testing and, 274–275
  prescriptions for, 279–280
Waste incineration by-products, 5, 28–29
Wastewater, reclaiming, 58
Water
  availability, threats to, 53–55
  contamination sources, 42
  infrastructure, 40
  quality (*see* Water quality)
  quantity, solutions for, 56–58
  resource conflicts, 55
  scarcity, health and, 88
  sources, 40
  treatment processes, 43–44, 55–56
  usage, increasing efficiency for, 57
  uses, 40–41
withdrawal, 57
Water-based disease, 44
Waterborne disease. *See also specific waterborne diseases*
  climate change and, 105, 106
  deaths from, 39
  in developing countries, 45
  diarrhea, 39, 46
  effect of water treatment on, 45
  transmission, 44
Water-dispersed infection, 44
Water quality
  biologic threats to, 44–47
  chemical threats to, 47–53
  industrial chemicals and, 163
  solutions for, 55–56
  threats to, 53–55
Water-related disease, 44
Water-washed disease, 44
Weather events, extreme, 104, 105
West Nile virus, 108
WHO. *See* World Health Organization
Wingspread Conference, 241
Winter, ozone depletion, 136–137
Women
  adverse health effects of background exposure, 186–189
  industrial chemical exposure, 184–185
Workers
  high-risk, 166
  occupational exposure of (*see* Occupational disease; Occupational exposure)
  personal hygiene of, 265
World Bank, 127
World Health Organization (WHO)
  air pollution reports, 16, 30
  arsenic drinking water standard, 51
  global diversity and, 127, 128
  health care report, 285
  waterborne disease transmission and, 44
World Resources Institute, 127
World Wide Fund for Nature, 289

Xenobiotics. *See also specific xenobiotics*
  body burden, 186
  definition of, 65
  metabolic models, 164–165
X-rays, 215–216

Ziconotide (SNX-111), 122–123
Zimbabwe, 88
Zinc, 2